Trailblazing Medicine
Sustaining Explorers During Interplanetary Missions

Erik Seedhouse

Trailblazing Medicine

Sustaining Explorers During Interplanetary Missions

Published in association with
Praxis Publishing
Chichester, UK

Erik Seedhouse, M.Med.Sc., Ph.D., FBIS
Milton
Ontario
Canada

SPRINGER–PRAXIS BOOKS IN SPACE EXPLORATION

ISBN 978-1-4419-7828-8   e-ISBN 978-1-4419-7829-5
DOI 10.1007/978-1-4419-7829-5

Springer New York Dordrecht Heidelberg London

Library of Congress Control Number: 2011921112

© Springer Science+Business Media, LLC 2011
All rights reserved. This work may not be translated or copied in whole or in part without the written permission of the publisher (Springer Science+Business Media, LLC, 233 Spring Street, New York, NY 10013, USA), except for brief excerpts in connection with reviews or scholarly analysis. Use in connection with any form of information storage and retrieval, electronic adaptation, computer software, or by similar or dissimilar methodology now known or hereafter developed is forbidden.
The use in this publication of trade names, trademarks, service marks, and similar terms, even if they are not identified as such, is not to be taken as an expression of opinion as to whether or not they are subject to proprietary rights.

Cover design: Jim Wilkie
Cover image: © E-spaces and Robert A. Freitas Jr, 3danimatione-spaces.com and www.rfreitas.com and Philippe van Nedervelde
Project copy editor: Christine Cressy
Typesetting: BookEns, Royston, Herts., UK

Printed on acid-free paper

Springer is part of Springer Science+Business Media (www.springer.com)

Contents

Preface ix
Acknowledgments xi
About the author xv
List of figures xvii
List of tables xix
List of panels xxi
List of abbreviations and acronyms xxiii

Section I Space Medicine 1

1   **Medicine onboard the International Space Station** 3
        Radiation 6
        Bone loss 9
        Muscle atrophy 10
        Behavioral health and fatigue 11
        Balance 12

2   **Interplanetary health care** 15
        Medical capabilities 15
        Medical strategies 19
        The interplanetary flight surgeon 20
        In-flight health care 27
        Anesthesia 32
        Airway management 32
        Surgery 33
        Rehabilitating astronauts 35
        References 37

3   **Medical qualification for exploration class missions** 39
        Medical selection of astronauts 39
        Genetic screening 47
            Types of genetic testing 49
        Precautionary surgery 49

	Appendicitis	52
	Appendectomy	52
	Pre-mission medical selection	53
	References	53

Section II Exploration Class Medical Challenges … 55

4 Radiation … 57
Radiation types … 57
Measuring radiation … 59
Radiation damage … 60
Radiation exposure guidelines … 64
Countermeasures … 66
    Operations … 66
    Shielding … 66
    Biological … 67
Nanotech … 68
    Installation … 73
References … 75

5 Bone loss … 77
Effect of microgravity on the skeletal system … 78
Countermeasures to bone demineralization … 82
Artificial gravity … 86
    Joosten's spinning spaceships … 89

6 Behavior and performance … 91
Shackleton … 94
Nansen … 96
Interplanetary stressors … 100
Mission operations … 109
Post-mission mental health care … 112

Section III Future Developments … 113

7 Bioethics, sex, and cloning … 115
The survivor scenario … 115
The terrorist attack scenario … 117
The injured crewmember scenario … 117
The solutions … 119
No sex please, we're astronauts … 121
Pulling the plug … 127
The one-way trip option … 127
Gattaca … 131
Cloning … 134
Cloning ethics … 138

8	**Robotic surgery and telemedicine**	141
	Telemedicine	141
	Supervisory-controlled systems	144
	Telesurgery	145
	Autonomous	148
	Nanobots	150
	Trauma pods	152
	Medical monitoring	155
9	**Stasis**	159
	Animal hibernation	162
	Human hibernation	165
	Entry	166
	Hibernation period	166
	Exit	167
	Stasis monitoring	170
	Life support	171

Appendix: The Interplanetary Bioethics Manual 173
Index 179

Preface

In 2009, with the International Space Station (ISS) declared fully operational, NASA and its partners ushered in a new era of spaceflight: permanent human presence in low Earth orbit (LEO). As the culmination of decades of manned spaceflight activities, the ISS focuses attention on what has been learned to date and on what must still be learned before humans can embark on future exploration endeavors. What we may discover during the forthcoming exploration of the solar system may shape the future of humanity, but before exploration class missions (ECMS) can leave LEO, we must be sure the astronauts will survive, which is why space medicine is the key to the future of humans in space.

Space medicine has undergone a gradual evolution, from developing and implementing selection and retention standards to minimizing the probability of disease in astronauts in space, to providing clinical support for short-duration missions and, most recently, to supporting a permanent human presence in space onboard the ISS. The ISS not only serves as an orbiting laboratory and technology development platform; it also provides clinicians with a unique opportunity to conduct research to optimize crew safety and performance – factors critical in reducing the biomedical risk of extended space missions.

Missions to Mars and beyond will test space medicine to the extreme. First, there is the question of how to protect astronauts from radiation that can pepper an astronaut's body like machine-gun fire. While mission planners are confident they can protect astronauts by using polyethylene shielding, there is a second kind of radiation for which there is no protection: cosmic rays possess too much energy for shielding to be effective. They pass through tissue, leaving cells mutated or dead, which means understanding their biological effects will be a priority. To protect themselves, astronauts may have to take anticancer drugs or, as suggested in this book, be infused with nanobots capable of repairing the damage inflicted by the radiation.

In addition to being fried by radiation, spacefarers embarking upon ECMs also face weakened muscles and significant bone loss. The obvious countermeasure to keep the muscles and bones fit is exercise, and the message to long-duration astronauts is clear: do the exercise and you will be okay – don't and you'll be carried off the spacecraft. However, even with rigorous daily exercise, astronauts still lose bone mass and scientists don't know when or if the body stops losing bone. If bone loss can't be prevented, there is the real risk that astronauts landing on some distant planet or moon will fracture bones. To counteract the bone loss, it has been suggested that astronauts take drugs normally given to osteoporosis patients, while another idea is to use artificial gravity to provide astronauts with doses of gravity to counteract the effects of weightlessness. These interventions and others, many of which are discussed in this book, will be vital in preparing for Mars missions and beyond. Chapter by chapter, this book examines the future of space medicine as it relates to human space exploration and describes what is necessary to keep a crew alive in space, how it is done today and how it will be accomplished in the future.

Acknowledgments

In writing this book, the author has been fortunate to have had five reviewers who made such positive comments concerning the content of this publication. He is also grateful to Maury Solomon at Springer and to Clive Horwood and his team at Praxis for guiding this book through the publication process. The author also gratefully acknowledges all those who gave permission to use many of the images in this book, especially scientists Dr Robert Freitas and Philippe van Nedervelde.

The author also expresses his deep appreciation to Christine Cressy, whose attention to detail and patience greatly facilitated the publication of this book, to Jim Wilkie for creating the cover of this book, and to Stewart Harrison, who sourced several of the references that appear in this book. Thanks also to Dr Gary Gray for his insight into the medical challenges of interplanetary missions.

Once again, no acknowledgment would be complete without special mention of our cats, Jasper, MiniMach, and Lava, who provided endless welcome (and occasionally unwelcome!) distraction and entertainment.

To Roald Amundsen, Sir Ernest Shackleton, Fridtjof Nansen, and the polar explorers who accepted the dangers and challenges of human endeavor and to the trailblazers of the future.

About the author

Erik Seedhouse is an aerospace scientist whose ambition has always been to work as an astronaut. After completing his first degree in Sports Science at Northumbria University, the author joined the legendary 2nd Battalion the Parachute Regiment, the world's most elite airborne regiment. During his time in the "Para's", Erik spent six months in Belize, where he was trained in the art of jungle warfare and conducted several border patrols along the Belize–Guatemala border. Later, he spent several months learning the intricacies of desert warfare on the Akamas Range in Cyprus. He made more than 30 jumps from a Hercules C130 aircraft, performed more than 200 abseils from a helicopter, and fired more light anti-tank weapons than he cares to remember!

Upon returning to the comparatively mundane world of academia, the author embarked upon a master's degree in Medical Science at Sheffield University. He supported his master's degree studies by winning prize money in 100 km ultradistance running races. Shortly after placing third in the World 100 km Championships in 1992 and setting the North American 100 km record, the author turned to ultradistance triathlon, winning the World Endurance Triathlon Championships in 1995 and 1996. For good measure, he also won the inaugural World Double Ironman Championships in 1995 and the infamous Decatriathlon, the world's longest triathlon, an event requiring competitors to swim 38 km, cycle 1,800 km, and run 422 km. Non-stop!

Returning to academia once again in 1996, Erik pursued his Ph.D. at the German Space Agency's Institute for Space Medicine. While conducting his Ph.D. studies, he still found time to win Ultraman Hawaii and the European Ultraman Championships as well as completing the Race Across America bike race. Due to his success as the world's leading ultradistance triathlete, Erik was featured in dozens of magazines and television interviews. In 1997, *GQ* magazine nominated him as the "Fittest Man in the World".

In 1999, Erik decided it was time to get a real job. He retired from being a professional triathlete and started his post-doctoral studies at Vancouver's Simon Fraser University's School of Kinesiology. In 2005, the author worked as an astronaut training consultant for Bigelow Aerospace in Las Vegas and wrote

About the author

Tourists in Space, a training manual for spaceflight participants. He is a Fellow of the British Interplanetary Society and a member of the Aerospace Medical Association. Recently, he was one of the final 30 candidates of the Canadian Space Agency's Astronaut Recruitment Campaign. Erik currently works as a manned spaceflight consultant, triathlon coach, and author. He is the Training Director for Astronauts for Hire (*www.astronautsforhire.org*) and plans to travel into space with one of the private spaceflight companies.

In addition to being a triathlete, sky-diver, pilot, and author, Erik is an avid scuba-diver and mountaineer and is currently pursuing his goal of climbing the Seven Summits. *Trailblazing Medicine* is his seventh book. When not writing, he spends as much time as possible in Kona on the Big Island of Hawaii and at his real home in Sandefjord, Norway. Erik lives with his wife and three rambunctious cats – Jasper, Mini-Mach, and Lava – on the Niagara Escarpment in Canada.

Figures

1.1	International Space Station	5
1.2	Galactic cosmic rays	6
1.3	Matroshka	7
1.4	Advanced Resistive Exercise Device (ARED)	10
2.1	A simple algorithm for tooth extraction	19
2.2	Kidney stones	23
2.3	Crewmember working in an extreme environment	27
2.4	Seawolf class submarine, USS *Connecticut* (SSN-22)	28
2.5	Zero-G intubation performed during parabolic flights conducted on an Airbus 300	33
3.1	Sir Ernest Shackleton	44
3.2	Lisa Nowak	46
3.3	Ethan Hawke in a scene from *Gattaca*	48
3.4	Leonid Rogozov performing his auto-appendectomy in the Antarctic	50
4.1	The Sun showing a C3-class solar flare, a solar tsunami, and multiple filaments of magnetism lifting off the stellar surface	59
4.2	A blood sample from an ISS astronaut that has been damaged by space radiation	62
4.3	Cell rover	69
4.4	Dendrimer complex docking on cellular folate receptors	70
4.5	Respirocyte in a blood vessel surrounded by red blood cells	71
5.1	Osteoporosis is an occupational hazard for long-duration astronauts	79
5.2	AMPDXA	81
5.3	Combined Operational Load-Bearing External Resistance Treadmill (COLBERT)	83
5.4	Sunita Williams	85
5.5	Artificial gravity	87
5.6	NASA's artificial gravity program	88
5.7	Promising research in the area of artificial gravity	90
6.1	MARS500 test crew	92
6.2	MARS500 facility	93

Figures

6.3	Shackleton's ship, the *Endurance*, trapped in ice	95
6.4	Launching the *James Caird* from the shore of Elephant Island, April 24th, 1916	96
6.5	Fridtjof Nansen	97
6.6	Nansen's ship, the *Fram*	98
6.7	The hut that Nansen and Johansen used as their winter quarters	99
6.8	Cramped living space during exploration class missions	101
6.9	Concordia research station	103
6.10	Shackleton's party, left behind on Elephant Island	105
6.11	The Skylab-4 crew	106
7.1	Artists rendition of Callisto	116
7.2	Entry, descent, and landing (EDL)	117
7.3	Jan Davis	123
7.4	Captain Oates	128
7.5	K2	131
7.6	Sam Bell (played by Sam Rockwell), from *Moon*	135
7.7	Sam Bell and his clone, from *Moon*	137
8.1	Telemedicine	142
8.2	Dr Mika Sinanan and Dr Thomas Lendvay collaboratively teleoperating Raven IV at the Bionicas Lab, UCSC	144
8.3	NASA Extreme Environment Mission Operations (NEEMO)	146
8.4	Aquarius undersea laboratory	147
8.5	Raven IV	148
8.6	Raven	149
8.7	Conceptual image of the Trauma Pod being lifted into the mobile unit	153
8.8	Conteptual Trauma Pod surgical tools poised for operation	154
8.9	The revolutionary Bio-Suit	156
9.1	Hypersleep, Hollywood style	160
9.2	Variable Specific Impulse Magnetoplasma Rocket (VASIMR)	162
9.3	Black bear	163
9.4	Arctic ground squirrel	164

Tables

1.1	Medical risks during exploration class missions.	4
2.1	Classification of illnesses and injuries in spaceflight	16
2.2	NASA medical training for International Space Station crewmembers	16
2.3	Exploration class mission medical supplies	17
2.4	Levels of prevention	20
2.5	In-flight medical events for US astronauts (STS-1 through STS-89)	29
2.6	Major health and medical issues during exploration class missions	29
2.7	Post-flight rehabilitation plan	35
3.1	Exploration class medical examinations and parameters	40
4.1	Short-term effects on humans of severe radiation	61
4.2	Radiation levels causing excess cancer risk	65
4.3	Chemical composition of candidate shielding materials	67
6.1	Behavioral stressors of long-duration spaceflight.	100
6.2	Comparison of psychologically relevant factors.	104
8.1	Concept of telemedical support.	143
9.1	Effect of stasis on the life-support-system requirements.	161
9.2	Physiological rates of black bears and arctic ground squirrels	165

Panels

1.1	Matroshka	8
2.1	Flight surgeon	21
2.2	Kidney stones	24
2.3	Toothache in orbit	31
2.4	Tissue-engineered Organ Replacement System (TORS)	34
3.1	Psychiatric assessment	44
3.2	Lisa Nowak	45
4.1	Solar flares	58
4.2	DNA mutation	63
4.3	Dendrimers	69
4.4	The vasculocyte	72
5.1	The COLBERT	84
7.1	Five Wishes	120
9.1	How squirrels hibernate	165

Abbreviations and acronyms

AI	Artificial Intelligence
ALARA	As Low As Reasonably Achievable
ALS	Advanced Life Support
AMPDXA	Advanced Multiple-Projection Dual-Energy X-ray Absorptiometry
ANARE	Australian National Antarctic Research Expedition
ARED	Advanced Resistive Exercise Device
ASCR	Astronaut Strength, Conditioning and Rehabilitation
ATP	Adenosine Triphosphate
BMD	Bone Mineral Density
BNL	Brookhaven National Laboratory
BNTR	Bi-modal Nuclear Thermal Rocket
CAD	Coronary Artery Disease
CAD-CAM	Computer-Assisted Design Computer-Assisted Manipulation
CAM	Centrifuge Accommodation Module
CCPK	Crew Contamination Protection Kit
CMC	Central Monitoring Computer
CMO	Crew Medical Officer
CMRS	Crew Medical Restraint System
CNS	Central Nervous System
COLBERT	Combined Operational Load-Bearing External Resistance Treadmill
CPD	Crew Passive Dosimeter
CRM	Crew Resource Management
CSA	Canadian Space Agency
CT	Computer Tomography
DARPA	Defense Advanced Research Projects Agency
DCS	Decompression Sickness
DEXA	Dual-Energy Absorptiometry
DMCF	Definitive Medical Care Facility
DNA	Deoxyribonucleic Acid
DSO	Defence Sciences Office
EBL	Exploration Bioethics Manual

Abbreviations and acronyms

ECH	Electrocardiogram
ECM	Exploration Class Mission
EDL	Entry, Descent, Landing
ENT	Ear, Nose, Throat
EPA	Eicosapentaenoic Acid
ESA	European Space Agency
ESM	Equivalent System Mass
ESWL	Extracorporeal Shock Wave Lithotripsy
EVA	Extra-vehicular Activity
FDA	Federal Drug Administration
Gy	Gray
GCR	Galactic Cosmic Ray
HCPOA	Health Care Power of Attorney
HIT	Hibernation Induction Trigger
HRF	Human Research Facility
ICRP	International Commission on Radiological Protection
ICU	Intensive Care Unit
IGF	Insulin-Growth Factor
IOP	Intraocular Pressure
ISEMSI	Isolation Study for European Manned Space Infrastructure
ISRU	In-Situ Resource Utilization
ISS	International Space Station
LEO	Low Earth Orbit
LET	Linear Energy Transfer
LOFT	Line-Oriented Flight Training
LSS	Life Support System
MCMI	Million Clinical Multiphasic Inventory
MCP	Mechanical Counterpressure
MEC	Medical Equipment Computer
MER	Medical Encounter Record
MIT	Massachusetts Institute of Technology
MMP	Mission Medical Pack
MMPI	Minnesota Multiphasic Personality Inventory
MRI	Magnetic Resonance Imaging
MSMB	Multilateral Space Medicine Board
NCRP	National Council on Radiation Protection
NEEMO	NASA Extreme Environment Mission Operations
NFKB	Nuclear Factor Kappa B
NIAC	NASA Institute for Advanced Concepts
NOAA	National Oceanic and Atmospheric Administration
NSBRI	National Space Biomedical Research Institute
NTDP	Nuclear Track Detector Package
PIS	Patient Imaging System
PKU	Phenylketonuria
PMC	Private Medical Conference

PNS	Peripheral Nervous System
PRS	Patient Registration Subsystem
RAM	Radiation Area Monitor
RBC	Red Blood Cell
RBE	Relative Biological Effectiveness
RAM	Radiation Area Monitor
SFINCSS	Simulation of Flight of International Crew on Space Station
SNS	Surgical Nurse Subsystem
SSD	Silicon Scintillator Device
SRAG	Space Radiation Analysis Group
SRC	Short Radius Centrifuge
SRL	Space Radiation Laboratory
SRS	Surgical Robot Subsystem
SRSF	Slow Release Sodium Fluoride
STP	Space Trauma Pod
TEPC	Tissue Equivalent Proportional Counter
TLD	Thermo-luminescence Dosimeter
TP	Trauma Pod
VASIMR	Variable Specific Impulse Magnetoplasma Rocket
VSS	Virtual Space Station

Section I

Space Medicine

1

Medicine onboard the International Space Station

Imagine the following scenario: the commander of Earth's first outer planets mission to Callisto is preparing to step onto the surface of the Jovian moon. Sleep-deprived, suffering from radiation sickness, weakened bones, and feeling discombobulated from months in zero gravity, she takes her first step on the icy surface and her femur snaps! She crashes to the surface and sustains a broken hip. The injuries render her helpless and she becomes a burden to the radiation-ravaged crew, who must provide around-the-clock medical attention. Stressed in their cramped spacecraft, which has served as home for more than two years, the crew bicker and squabble among themselves before venting their frustrations on Mission Control back on Earth. Fox News sensationalizes the problems, saying the crew has decided to euthanize the commander – something the space agency's public relations office vehemently denies. Attempts to stabilize the situation fail and the mission is threatened. The follow-up mission is cancelled.

It's a worst-case scenario, but it's entirely plausible, especially when one understands the myriad medical challenges (Table 1.1) that will be faced by interplanetary explorers. The longer astronauts are away from home, the greater the risks and the more dependent they will become on clinical care based on a thorough understanding of diagnosis and therapy of illness and injury in space. Space medicine is currently entering an evolutionary phase of incorporating near and over-the-horizon medical care capabilities that will be required when astronauts embark upon exploration class missions (ECMs) beyond low Earth orbit (LEO). Serving as a test-bed to evaluate some of these future medical capabilities is the International Space Station (ISS), which functions as a high-fidelity platform for assessing everything from bone loss countermeasures to operational health and performance.

The ISS (Figure 1.1) has maintained an uninterrupted human presence in space since the launch of Expedition 1 on October 31st, 2000. In fact, by the time you read this, the program will have exceeded the record set aboard the Russian space station Mir of 3,644 days (8 days short of 10 years). Designed primarily as a research laboratory, the ISS offers an advantage over spacecraft such as the Space Shuttle because it is a long-term platform ideally suited for resolving the many problems faced by future interplanetary crews. In fact, it is likely many of the medical systems,

Table 1.1. Medical risks during exploration class missions.

System	Risk	Description
Bone	Bone loss and fracture	Failure to recover bone during mission places crewmembers at risk of fracture upon landing
Bone	Impaired fracture healing	Fractures occurring during and immediately following long-duration spaceflight will require prolonged period for healing. Bone may not be completely restored due to changes in bone metabolism during flight
Bone	Renal stone formation	Urine calcium concentration is increased due to increased bone resorption
Cardio	Dysrhythmias	May cause low blood pressure and syncope (fainting)
Cardio	Impaired cardiac function	Long-duration spaceflight may result in a decrease in cardiac mass and result in altered cardiac function that could be irreversible
Immune	Infection	Long-duration spaceflight may depress the immune system and result in a greater number of infections
Immune	Allergies	Failure of the immune system may cause immunologic disease
Muscle	Atrophy	Muscles waste away during spaceflight, resulting in reduced muscle-force contraction and compromised movement skills
Muscle	Muscle damage	Atrophied muscles result in increased susceptibility to damage and soreness
Neuro	Vertigo and illusions	Transition between gravity environments (such as from a microgravity environment to a surface environment) may cause spatial disorientation and vertigo
Neuro	Balance	Adapting to a gravitational environment after spending several months in space may disrupt balance and locomotion
Clinical	Illness and trauma	Interplanetary astronauts will have no abort capability if injured or ill. Lack of capability to treat these injuries/illnesses may pose a threat to the mission
Clinical	Pharmacology delivery	Administration of drugs may be altered in microgravity. It may be impossible to treat some medical conditions, resulting in a threat to life and mission
Behavioral	Human performance	Poor interpersonal dynamics and team cohesiveness may compromise human performance and threaten mission success
Behavioral	Fatigue	Long-duration spaceflight missions are emotionally and physically exhausting. Circadian patterns are disrupted and mission demands and timelines result in long work hours. Human error may occur when performing critical tasks
Radiation	Biological effects	Heavy iron particles striking the brain can impair motor ability, cognition, and memory
Radiation	Radiation sickness	Cosmic rays may leave cells unstable, mutated, or dead. High radiation doses may result in radiation sickness and death

Medicine onboard the International Space Station 5

Figure 1.1 The International Space Station has served as a test-bed for developing the medical capabilities that will be required for exploration class missions. Image courtesy: NASA.

technologies, and protocols incorporated into future ECMs will be thanks to the research performed onboard the ISS. With that in mind, it is worth discussing some of the research space agencies are conducting onboard the orbiting laboratory and how these studies will help astronauts venturing beyond LEO.

Most medical experts and mission planners agree the biggest risks faced by interplanetary explorers are those posed by radiation exposure, bone loss, and the effect on behavioral health. Of these three, radiation exposure probably has flight surgeons the most worried. While astronauts working onboard the ISS in LEO are relatively protected, crewmembers venturing into deep space will face an onslaught of radiation. The most dangerous kind of radiation ECM astronauts will experience is from galactic cosmic rays (GCR), bare atomic nuclei, some as heavy as iron atoms, accelerated to nearly the speed of light by distant supernovas. Because of their high velocity, high mass, and positive electric charge, GCR (Figure 1.2) particles can cause tremendous damage to an astronaut's cells.

Figure 1.2 Galactic cosmic rays can inflict serious damage upon crewmembers. Image courtesy: National Oceanic and Atmospheric Administration.

RADIATION

To investigate the radiation environment onboard the ISS, researchers designed a human-shaped torso and strapped it outside the station. Named Matroshka (Figure 1.3), the torso is an armless, legless mannequin that looks like it's wrapped in a mummy's bandages. It also happens to be an intrepid space traveler, as it spent several months on the ISS helping scientists learn how they can best protect future interplanetary astronauts from the effects of radiation. Before Matroshka (nicknamed the Phantom Torso, for obvious reasons), scientists were only able to estimate radiation doses using computer models, but computer models and real life are often very different. Until Matroshka (Panel 1.1) came along, researchers weren't sure whether their models accurately predicted the radiation dose astronauts experience in space.

You see, what really matters is how much radiation actually hits an astronaut's vital organs. To reach those organs, radiation particles must first pass through the walls of the spacecraft, the astronaut's spacesuit, and, finally, their skin and other body tissues. Sometimes, these barriers will slow down or stop a radiation particle, but sometimes the collision between a radiation particle and a barrier will produce a

Radiation 7

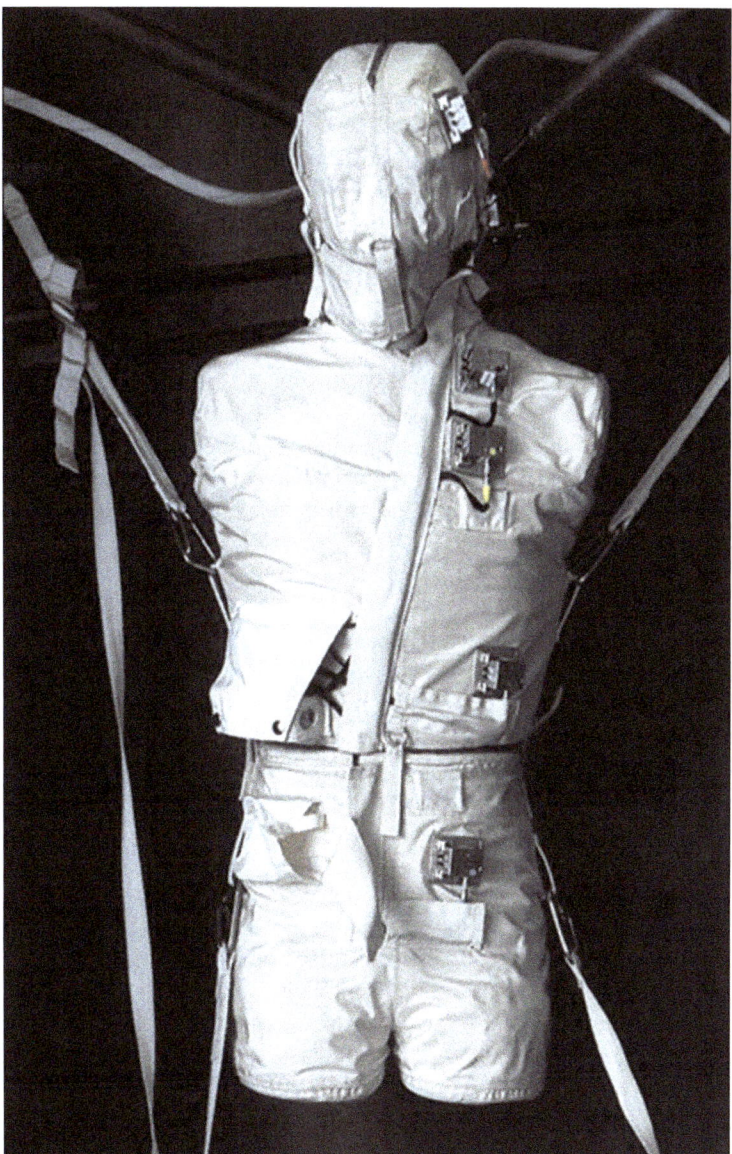

Figure 1.3 Matroshka was sent to the International Space Station aboard the Soyuz-U/Progress M1-11 supply vehicle launched on January 24th, 2004, and was placed on the outside of the Russian Zvezda module during a spacewalk performed by Alexander Kaleri and Michael Foale on March 15th, 2004. It was brought inside on August 18th, 2005, and the experimental elements were returned to Earth on October 11th, 2005. Matroshka received more experimental elements on December 21st, 2005. These "slices" measured similar data for conditions inside the ISS until active detectors were received later to continue dose readings until its return to Earth. The torso was returned to Earth in 2009. Image courtesy: NASA.

> **Panel 1.1. Matroshka**
>
> Matroshka consisted of commercial parts, common to the field of radiotherapy. It had several slices composed of natural bones, embedded in plastics simulating different tissue densities. The torso slices were equipped with channels allowing special sensors to be accommodated. Each Matroshka slice had a center hole that allowed the assembly of the whole torso to fit over a rod that was fixed to the base structure. Detector systems were located at the sites of organs that are the most radiation-sensitive parts of the body. The radiation measurement instruments included five nuclear track detector packages (NTDP), several thermo-luminescence dosimeter (TLD) packages, and five small silicon scintillator devices (SSDs), which were mounted inside the torso. These instruments allowed scientists to measure the absorbed dose, neutron dose, and the heavy ion fluences, thereby helping scientists update computer models and develop new models that will be used on flights beyond LEO.

shower of new radiation particles, called "secondary radiation". Although ISS crewmembers wear sensors on their flight suits to record total radiation exposure, there's no practical way to measure just how much radiation actually reaches their vital organs, which is why Matroshka was so useful. It was made of a special plastic that closely mimics the density of the human body, sliced horizontally into 35 layers. In these layers, researchers embedded hundreds of lithium-crystal dosimeters (radiation detectors), each capable of measuring the accumulated radiation dose at one point in the body. The dosimeters, which were located where vital organs such as their brain, thyroid, heart, colon, and stomach would be, kept a record of how the radiation dose changed moment by moment. Together, these sensors documented how radiation propagated through astronauts' bodies and provided the real-world test needed to prove that the computer models were mostly correct. In fact, by analyzing the measurements from hundreds of radiation sensors embedded throughout the torso's body, NASA scientists found that the computer models were accurate to within 10% of the measured dose.

But, even with Matroshka and the ongoing radiation research being performed on the ISS, scientists acknowledge they have a long way to go before they can keep interplanetary astronauts safe. In an attempt to shield today's crews, selected areas of the ISS have enhanced radiation shielding (polyethylene foam and water are the materials that currently provide the best protection from space radiation), affording more protection to crewmembers when needed. There are also plans to use polyethylene in the sleep stations. But, while these procedures may help protect astronauts in LEO, they won't be sufficient to shield astronauts traveling in deep space, which is why new technologies, such as those described in Chapter 4, will be required.

BONE LOSS

After so many long-duration missions in LEO, it is now an established fact that astronauts spending months in space lose significant bone strength, making them increasingly at risk for fractures later in life. Studies performed by California universities that evaluated 13 astronauts who spent four to six months on the ISS found that, on average, astronauts' hip bone strength decreased 14%. In fact, three astronauts experienced losses of 20–30% – rates comparable to those seen in older women with osteoporosis. For mission planners and astronauts contemplating multi-year missions, the rate of loss is worrying because if preventive measures aren't taken, some astronauts may be at increased risk for fractures as soon as they step out of the spacecraft.

For as long as there have been astronauts, scientists have studied why the microgravity environment makes bones so fragile, but the California studies were the first to evaluate bone strength. The researchers used a special computer program to identify hip bone fracture risk in people with osteoporosis. The program, which was used to analyze the hip bone scans of one female and 12 male ISS crewmembers, measured a decrease in bone strength of between 0.6 and 5.0% for each month of duty onboard the station. For astronauts embarked upon a multi-year mission to Callisto or some other far-flung destination, such a rate of loss could mean a death sentence (remember the scenario at the beginning of this chapter?). Here on Earth, a hip fracture almost always requires hospitalization and major surgery. It can impair a person's ability to walk unassisted and may cause prolonged or permanent disability or even death. For an astronaut to suffer such an injury a billion kilometers from home is a scenario that doesn't bear thinking about.

While the California study focused on the decrease in bone strength, other studies have documented the loss of bone mass in crewmembers. One study spanning four years found that ISS crewmembers lost about 11% of their total hip bone mass during the course of their mission. Such a rate of loss equates to losing as much bone mass in one month in LEO as an elderly woman loses in an entire year! Although the study found that a year after the crewmembers had returned to Earth, much of the lost bone mass had been replaced, the bone structure and density had not returned to normal and hip strength had not recovered; astronauts landing on a distant planet won't have the luxury of waiting for their bones to recover.

The problem of bone loss is compounded by the loss of calcium that leaches out of the bones, which puts astronauts at risk of kidney stones. At least a dozen American crewmembers have developed kidney stones in the last five years, and as interplanetary missions loom on the horizon, the number is likely to grow. It's a real health concern for astronauts because kidney stones are mineral deposits in the kidneys that can travel through the urinary tract, causing intense pain. In fact, the pain is so intense that if an astronaut were to suffer the condition onboard the ISS, it is likely the crewmember would be returned to Earth. En route to Mars, a mission abort won't be an option.

10 Medicine onboard the International Space Station

Figure 1.4 The Advanced Resistive Exercise Device (ARED) is a versatile machine that can be used to perform all sorts of exercises. Conceived as a means of helping astronauts to maintain muscle and bone strength and endurance in low-gravity environments, the ARED allows astronauts to perform resistive exercises for stimulating bone regeneration and exercising the major muscle groups. Image courtesy: NASA.

MUSCLE ATROPHY

While radiation exposure and bone loss are major causes for concern, mission planners are also uneasy about the muscle wasting that occurs during long sojourns in space. Flight surgeons have seen the muscles that astronauts use to stand and walk begin to waste away after only a few weeks in LEO, so it's obvious a multi-year mission poses a huge risk. The reason for scientists being so interested in exercise is because loss of muscle mass and strength continues throughout the mission, even if crewmembers stick to a strict exercise program. For example, in one study, scientists compared the power output between astronauts who exercised more than 200 min per week with the power output of those that exercised less than 100 min per week. While the astronauts who exercised longer performed better than those who exercised least, the extra exercise only slowed the atrophy of their muscles.

Onboard the ISS, astronauts work out on all sorts of exercise equipment in an attempt to preserve their muscle strength. For example, crewmembers perform resistive exercise using the Advanced Resistive Exercise Device – a strength device (Figure 1.4) that imparts load on the body.

In addition to using the ARED, astronauts spend much of their time pedaling a recumbent cycle, which provides general aerobic and cardiovascular conditioning as well as improved muscular endurance. Cycle ergometry is also an important aspect of physical conditioning for space walks and during the pre-breathe exercise period before a space walk. Astronauts perform the ergometry either in the manual mode, with the workload controlled manually by the astronaut, or in electronic mode, with the workload varied by an electronic controller.

Now, you might think resistive exercise and cycling would be enough to keep muscles healthy, but it isn't. Astronauts also use a Combined Operational Load-Bearing External Resistance Treadmill (COLBERT in NASA parlance) treadmill that provides medical and science personnel with more physiological data than ever before on astronauts' exercise capabilities. While this information will be useful for planning exercise schedules, scientists know that it will be of limited value when crews are embarked upon missions lasting several years, which is why interplanetary spaceships will most likely be fitted with some sort of artificial gravity capability. It's a subject that's discussed in Chapter 5.

BEHAVIORAL HEALTH AND FATIGUE

Following radiation and bone loss, perhaps the next greatest concern among mission planners is behavioral health. Behavioral health risk increases with mission duration, which is why preventive measures begin during mission training by preparing ISS crewmembers for the environment they will experience. For example, private medical and behavioral health conferences between the crewmember and specially trained ground medical staff are held regularly. Robust family support is routine and includes regular video conferences between the crewmember and the family. Additionally, e-mail and private telephone calls are available for the crewmembers to communicate with their friends and family on Earth. The use of such preventive measures will help mission planners answer questions such as "What would happen to a person's mental and behavioral health if he or she were cooped up with four or five other individuals during a three-year period?" and "What does living and working together in such close quarters over such an extended time do to group interactions?" Now, you might think behavioral problems would be the least of mission planners' worries. After all, astronauts are rigorously screened for that sort of thing, aren't they? Well, yes they are, but as experience in the Antarctic has shown, schisms, friction, withdrawal, competitiveness, scapegoating, and other maladaptive behaviors are found even among highly competent men and women working together. Anyone remember Skylab-4? Every day, Mission Control sent the three astronauts a six-foot-long sheet of instructions. It was NASA's attempt to try and squeeze every minute out of the astronauts' days. The agency even tried scheduling experiments during the crew's mealtimes. Finally, and predictably, the astronauts snapped. In a stinging reprimand to Mission Control, Commander Gerald Carr informed the ground that the crew was on strike. After the astronauts spent some time relaxing and generally doing as they pleased, ground controllers finally got the message that astronauts needed time off.

In addition to observing astronaut behavior onboard the ISS, scientists are also interested in monitoring behavioral health from Earth, and what tools might be developed so that flight crewmembers can monitor their own moods and cognitive functioning. For example, scientists are evaluating a portable brain scanner (discussed in Chapter 6) designed to remotely monitor astronauts for signs of brain injury, depression, and even mental fatigue that might compromise a crewmember's ability to conduct mission-critical tasks. The scanner, which resembles a large remote control tethered to a Velcro headband by long, thin wires, works like a breathalyzer for the brain, and uses near-infrared optical spectroscopy to measure changes in blood flow to the brain. The system relies on an optical scanner that sends weak pulses of near-infrared light into the brain, then reads back the reflected wavelengths, which reveals how much oxygen is in the blood and, in turn, provides a measure of brain activity. Once deployed on the ISS, the scanner may look for changes in brain activity in regions that have been previously linked to depression, or it could be used to sense brain damage caused by environmental problems such as low oxygen or carbon monoxide levels in the station. During an interplanetary mission, such a device may help crews avoid close calls by detecting signs of mental stress before they're apparent to an astronaut or the crew. However, such a sensor may not be popular with astronauts, who may not take kindly to having a little black box telling them what they can and cannot do!

Another aspect of human performance scientists evaluate onboard the ISS is crew fatigue. Long-duration spaceflight missions are emotionally and physically exhausting. Not only are normal circadian patterns disrupted by endless sunrises and sunsets (every 90-min orbit), but mission demands and timelines result in long work hours, so it's not surprising that many ISS crewmembers report a lack of restful sleep. In fact, degradation in performance of tired crews has been considered comparable to the degrading effects of alcohol ingestion. Already, these effects have affected orbital operations, with one ISS crewmember stating that "We were falling asleep while repositioning the Soyuz". Once again, the experiences of ISS crews are proving invaluable for those tasked with defining work and rest schedules for longer-duration missions. For example, flight rules and planning constraints have recently been put in place to mitigate fatigue risk and critical operations during circadian lows, and sleep shifting has been implemented to accommodate complex operations.

BALANCE

Assuming scientists resolve the problems of radiation exposure and bone loss and assuming astronauts can endure several months or years locked inside a vehicle staring at the same faces without going crazy, there is still the problem of balance to contend with once they finally arrive at their destination. You see, astronauts returning from long-duration missions routinely face the challenge of simply standing up and walking. Imagine landing on Mars and not being able to walk in a straight line! You may laugh, but it's a very real risk. These balance disturbances cause astronauts to suffer from dizziness and mean they have difficulty standing, walking, and turning corners. To perform everyday activities, the brain interprets information provided by the body's sensory systems, which include the eyes, the inner-ear balance organs, and muscle-movement receptors. The problem for long-duration astronauts occurs during the transition period when the brain is trying to adapt to a new gravity environment, which could be the 1-G environment of Earth or adjusting to the one-third gravity of Mars. You see, when you're in space, information from the sensory systems is different, forcing the brain to reinterpret information and make adjustments to allow you to do the activities you need to do in space. The problem is when astronauts return to Earth (or land on a moon or planet after a long journey) because the sensory systems aren't used to normal gravity. Some astronauts returning from six-month increments onboard the ISS have compared the discombobulating balance disturbances to those experienced after stepping off a fast-spinning merry-go-round. But, while the effects disappear after a merry-go-round ride, the effects following a spaceflight are far more persistent, with some symptoms lasting for weeks. It's one of the reasons some astronauts wear their NASA issue diapers (and this is according to a shuttle pilot I spoke with) to bed during the first few days back on Earth, in case they feel the urge to visit the bathroom and can't make it without falling over. Making matters worse is the fact that these disturbances aren't limited to just standing up and walking. Some people with inner-ear trouble don't steer vehicles very well, which doesn't bode well for an early start to surface operations. In fact, some researchers think astronauts may have to spend some time adapting to the gravity environment before setting foot on the surface of a new planet.

Research onboard the ISS focuses our attention on what we have learned to date and what must still be learned before embarking upon ECMs. Space medicine has been an integral part of success in the manned spaceflight arena and will play a critical role when crews finally leave LEO. To prepare for that day, space medicine experts will develop new and, in some cases, radical technologies to protect crews against the medical risks faced en route to destinations beyond the orbit of the ISS. Today, space agencies can ensure the health and safety of a crew in LEO. Tomorrow, they must be able to ensure the health and safety of a crew millions of kilometers away from Earth. This will require self-repairing systems, autonomous health care applications and myriad new technologies, some of which are only just being developed. Long-duration missions onboard the ISS will continue to yield space medicine data, validate concepts, test hypotheses, and develop countermeasures, many of which are outlined in this book.

2

Interplanetary health care

Exploration class missions (ECMs) will expose astronauts to several unique and hazardous elements. The isolation and great distances mean evacuation will not be an option and the crew and the flight surgeon will need to be prepared to deal with myriad medical situations ranging from motion sickness to death. Compounding the problem of responding to diverse medical situations will be the limited medical capabilities onboard, the cramped living and working quarters, and, of course, the challenges of performing medical procedures in the microgravity environment. Given the limitations of deep space health care, it will be more important than ever to implement effective medical mitigation strategies and new flight surgeon training methods. Here, in Chapter 2, ECM medical capabilities and strategies and the future role of the flight surgeon are discussed.

MEDICAL CAPABILITIES

In February 2008, an undisclosed medical issue among the crew of the Space Shuttle *Atlantis* prompted a 24-hr delay to a scheduled extravehicular activity (EVA). ESA astronaut, Hans Schlegel, was eventually replaced by NASA astronaut, Stanley Love, and later rejoined the EVA rotation. The incident was typical of the many minor medical conditions astronauts suffer during spaceflight. To date, the spectrum of medical conditions (Table 2.1) reported by NASA and ESA astronauts have rarely required serious medical attention and there has been no medical evacuation of any NASA or ESA crewmember. However, given the extreme nature of the space environment combined with the duration of an ECM, it is inevitable that sooner or later, medical intervention will be required.

Since it is not certain that every mission will have a physician-astronaut, the burden of any ECM medical contingency will fall upon the shoulders of the crew medical officer (CMO). At present, the CMO is a pilot or scientist with 34 hr of medical training, whereas other crewmembers receive only 17 hr of pre-flight medical training. However, for ECMs lasting several years, crew medical training may be increased and astronauts selected for these missions will probably follow a schedule similar to that outlined in Table 2.2.

Table 2.1. Classification of illnesses and injuries in spaceflight.

Characteristics	Examples	Type of response
Class I		
Mild symptoms	Space motion sickness	Self care
Minimum effect upon performance	Gastrointestinal distress	Administration of prescription and/or non-prescription medication
Non-life-threatening	Urinary tract infection	
	Upper respiratory infection	
	Sinusitis	
Class II		
Moderate to pronounced symptoms	Decompression sickness	Immediate in-flight diagnosis and treatment
	Air embolism	
Significant effect upon performance	Cardiac arrhythmia	Possible evacuation
	Toxic substance exposure	Possible mission termination
Potentially life-threatening	Open/closed chest injury	
	Fracture	
	Laceration	
Class III		
Immediate severe symptoms	Explosive decompression	Immediate evacuation following resuscitation and stabilization if necessary
Incapacitating	Overwhelming infection	
	Massive crush injury	
Unsurvivable if definitive care unavailable	Open brain injury	Comfort measures applied
	Severe radiation exposure	

Table 2.2. NASA medical training for International Space Station crewmembers [1].

Training session	Crew	Time	Time prior to launch
ISS space medicine overview	Entire crew	0.5 hr	18 months
Crew health care system (CHeCS) overview	Entire crew	2 hr	18 months
Cross-cultural factors	Entire crew	3 hr	18 months
Psychological support familiarization	Entire crew	1 hr	18 months
Countermeasures system operations 1	Entire crew	2 hr	12 months
Countermeasures system operations 2	Entire crew	2 hr	12 months
Toxicology overview	Entire crew	2 hr	12 months
Environmental health system microbiology operations and interpretation	ECLSS	2 hr	12 months
Environmental health system water quality operations	ECLSS	2 hr	12 months
Environmental health system toxicology operations	ECLSS	2 hr	12 months
Environmental health system radiation operations	ECLSS	1.5 hr	12 months
Carbon dioxide exposure training	Entire crew	1 hr	12 months
Psychological factors	Entire crew	1 hr	12 months
Dental procedures	CMOs	1 hr	8 months
ISS Medical diagnostics 1	CMOs	3 hr	8 months
ISS Medical diagnostics 2	CMOs	2 hr	8 months

ISS Medical therapeutics 1	CMOs	3 hr	8 months
ISS Medical therapeutics 2	CMOs	3 hr	6 months
Advanced cardiac life-support (ACLS) equipment	CMOs	3 hr	6 months
ACLS pharmacology	CMOs	3 hr	4 months
ACLS protocols 1	CMOs	2 hr	4 months
ACLS protocols 2	CMOs	2 hr	4 months
Cardiopulmonary resuscitation	Entire crew	2 hr	4 months
Psychiatric issues	Entire crew	2 hr	4 months
Countermeasures system evaluation operations	CMOs	3 hr	4 months
Neurocognitive assessment software	Entire crew	1 hr	4 months
Countermeasures system maintenance	Entire crew	2.5 hr	4 months
Environmental health system Preventive and Corrective Maintenance	Entire crew	1 hr	4 months
ACLS "megacode" practical exercise	Entire crew	3 hr	3 months
Psychological factors 2	Entire crew	2 hr	1 month
Medical refresher	Entire crew	1 hr	2 weeks
CMO computer-based training	CMOs	1 hr/month	During mission
CHeCS health maintenance system contingency drill	Entire crew	1 hr	During mission

To ensure adequate treatment and rehabilitation during ECMs, space agencies will also rely on instructing the crew using curricula and algorithms such as the one shown in Figure 2.1. In addition to extensive medical training, ECM crews will require a versatile medical system (Table 2.3) designed to stabilize and treat crewmembers. While an astronaut requiring urgent medical attention onboard the ISS can be evacuated to a definitive medical care facility (DMCF) on Earth, an injured crewmember en route to Mars won't have that option. For this reason, the provision of a medical system will present unique challenges to mission planners but Table 2.3 gives you some idea of what the onboard medical inventory might consist of.

Table 2.3. Exploration class mission medical supplies.

Basic medical system
Airway and trauma sub-pack (resuscitator and valve mask)	Operational bioinstrumentation system to provide downlink
EENT sub-pack (diagnostic items)	Patient and rescuer restraints
Drug sub-pack (oral and injectable)	Contaminant clean-up kit
Saline bag	Medical accessory kit
Intravenous administration sub-pack	Sharps container
Pharmaceutical sub-pack	Body bags

Extended medical system

Advanced Life Support (ALS)
Injectable medications	Blood pressure cuff
Intravenous fluid and administration equipment	Stethoscope
Airway management equipment	Pulse oximeter

18 **Interplanetary health care**

Mission medical pack

Oral medications	Portable blood analyzer
Topical medications	Dental hardware
Bandages	Minor surgical supplies
Banked synthetic blood and blood marrow	

Systems

System/kit	Description
Crew Medical Restraint System (CMRS)	Provides spinal stabilization of an injured crewmember and provides restraint for CMO treating the patient
Crew Contamination Protection Kit (CCPK)	The CCPK is a multipurpose clean-up kit that protects astronauts from toxic and non-toxic particles
Medical Equipment Computer (MEC)	The MEC is a versatile laptop that is the heart of the medical capability onboard: Displays physiological data from exercise devices Collects and stores medical data Maintains health records Assesses crew health Provides up/downlink capability Stores templates for custom-organ generation
Defibrillator	Also provides heart rate monitoring and analysis
Respiratory Support Pack	This system ventilates an unconscious astronaut automatically and provides oxygen to a conscious crewmember
Blood pressure/ECG	Measures systolic and diastolic blood pressure and also monitors and displays heart rate/ECG during exercise
Countermeasures System	Treadmill, resistive exercise device and bike ergometer
Magnetic Resonance Imaging System (MRIS)	The MRIS is a medical imaging technique used in radiology to visualize the detailed internal structure and function of the body
Autonomous Surgical Robot System (ASRS)	System capable of performing a variety of surgical interventional tasks, ranging from lesion biopsies to foreign body removal
Medical Telepresence System (MTS)	To perform surgery using telepresence, some devices, known as the teleoperated devices, are placed into the patients' internal organs to be operated. Using the MTS, the surgeons manipulate these instruments to see what is happening using small cameras located at the work site
Tissue-engineered Organ Replacement System (TORS)	Engineered biological tissues are customizable and immune-compatible (e.g. heart, limbs, eyes, lungs, pancreas, and bladder)
Interplanetary Bioethics Manual	To provide mission commanders and CMOs guidance with quandaries such as terminally ill crewmembers

DENTAL - TOOTH EXTRACTION
(ISS MED/3A - ALL/FIN) Page 1 of 3 pages

> **NOTE**
> Tooth Extraction is a last resort and is reserved only for those cases where pain is excessive or an infective process has set in and the amount of time remaining for the mission is greater than the time to safely control infection with antibiotics. A course of antibiotics will not cure a tooth infection, and more definitive care is always necessary. Extraction should only be done when all other treatment options have been exhausted and on consultation with Surgeon.

AMP
(blue)

1. Unstow from Dental Subpack:
   Elevator, 301 (Dental-4)
   34S (Dental-4)

> **NOTE**
> Type number is engraved on probe.

Gauze Pads (4) (P3-B4)
and one of following:

   Forceps, 151AS (Dental-3)
     (for incisors, cuspids, bicuspids)
   Forceps, 17 (for lower molars) (Dental-3)
   Forceps, 10S (upper molars) (Dental-3)

> **NOTE**
> Type number is engraved on probe.

2. Anesthetize area where tooth is to be extracted.
   Refer to {DENTAL - INJECTION TECHNIQUE} (SODF: ISS MED: DENTAL).

Figure 2.1 A simple algorithm for tooth extraction. Image courtesy: NASA.

MEDICAL STRATEGIES

Although the crew may be many million kilometers from home, taking care of an injured or ill astronaut during an interplanetary mission will follow the accepted medical strategy of primary, secondary, and tertiary prevention (Table 2.4).

While primary and secondary levels of care won't present too many problems, providing tertiary-level care will be challenging. A loss of pressure in a spacesuit during a spacewalk could result in a nasty injury known as spontaneous pneumothorax. If this were to happen during an ISS mission, it would be considered a mission-terminating event, but en route to Mars, it would require the insertion of a chest tube by the CMO. Another possible event is decompression sickness (DCS), which could easily incapacitate a crewmember. Once again, the CMO would need to

Table 2.4. Levels of prevention.

Prevention	Definition	Rationale	Methods
Primary	Measures implemented to avoid disease/illness	Eliminate hazard by selecting astronauts without disease	Estimating the incidence of disease in the astronaut corps. Genetic screening
Secondary	Measures aimed at identifying disease, thereby increasing opportunities for interventions to prevent progression of the disease	Protect against a risk that couldn't be controlled by primary prevention (e.g. bone loss)	Countermeasures such as load-bearing exercise and artificial gravity
Tertiary	Measures that reduce the negative impact of an established disease by restoring function and reducing complications	This level is implemented when the first two levels have failed	Injury/illness may result due to an uncontrolled event such as decompression. This is when Advanced Life Support (ALS) would be used

implement tertiary care, possibly by using Advanced Life Support (ALS) techniques. A third possible event is a crewmember suffering acute radiation sickness (discussed in more detail in Chapter 4). Obviously, a high level of skill will be required by those charged with providing tertiary care, especially when it will not be possible to return a crewmember to a DMCF. Not only will interplanetary CMOs have to deal with the challenges of administering ALS with limited resources, but they will also have to face the very real possibility of treating illnesses and injuries from which the crewmember may not survive. In fact, extended treatments for a severely injured crewmember could deplete irreplaceable consumables and jeopardize other crewmembers. In such a case, it may be the CMO's call to "treat with final resolution" and euthanize the astronaut. Needless to say, the job of CMO won't be an easy one.

THE INTERPLANETARY FLIGHT SURGEON

Flight surgeons (Panel 2.1) often say they have the second best job in the space agency because in terms of mission operations, they participate in many, if not all, of the same training as crewmembers. In addition to supporting pre-flight crew selection and training, monitoring on-orbit crew, and supervising post-flight rehabilitation, flight surgeons also evaluate medicines for spaceflight, and talk to the families of crewmembers on orbit. It's a demanding job, the most challenging aspect of which is trying to balance the responsibilities of serving the agency and the patient, in whom millions of dollars have been invested. For example, a major medical event during a mission may not only disrupt work schedules and create

The interplanetary flight surgeon 21

> **Panel 2.1.** Flight surgeon
>
> When most people think of flight surgeons, they think of Dr "Bones" McCoy of *Star Trek* fame, but flight surgeons don't fly in space unless they become astronauts. Take Michael Barratt, for example. Barratt is a doctor-turned-astronaut who spent more than six months as a real-life Dr McCoy onboard the ISS as the station's CMO. A veteran mountaineer and diver, Barratt spent nine years as a ground-based flight surgeon before deciding he wanted to experience the effects of spaceflight for himself. He became an astronaut in 2000 at the age of 41, but had to wait five years before being assigned to long-duration flight training and another four years before his first mission. He finally launched on Soyuz TMA-14 on March 26th, 2009, to the ISS and served as a member of Expeditions 19 and 20. After logging 199 days on orbit, Barratt landed on October 11th, 2009.

family stress, but it may also jeopardize the lives of other crewmembers who may be involved in a de-orbit evacuation, and create political strain between international partners.

Like any occupation, the job of flight surgeon comes with its fair share of administration. For example, flight surgeons meet weekly with medical boards to discuss the flight status of astronauts, they confer with representatives of other space agencies, present papers at conferences, and, more recently, involve themselves in the issues facing the next generation of space explorers.

While the medical challenges faced by planetary-bound astronauts will be much more hazardous than those faced by those confined to low Earth orbit (LEO), the manner in which problems will be resolved will follow a similar pattern. For example, during ISS missions, flight surgeons work in Mission Control and hold daily conferences (although these conferences are private, if the flight surgeon learns of something that might affect the mission, they let the flight director know) with the astronauts. During ECMs, it is likely the CMO will be a flight surgeon who will be in daily contact with flight surgeons in Mission Control. Perhaps the best way to understand what these flight surgeons might do during an interplanetary mission is to imagine a hypothetical medical event during an ECM.

We'll imagine you're a flight surgeon working at Mission Control during an interplanetary mission to Callisto,[1] the fourth of Jupiter's Galilean moons. We'll

---

[1] A 2003 NASA-led study identified revolutionary concepts and supporting technologies for Human Outer Planet Exploration (HOPE). Callisto, the fourth of Jupiter's Galilean moons, was chosen as the destination for the HOPE study. Assumptions for the Callisto mission included a launch year of 2045 or later, a spacecraft capable of transporting humans to and from Callisto in less than five years, and a requirement to support three humans on the surface for a minimum of 30 days.

assume you're certified in a variety of aerospace medical disciplines and have received extensive training that qualifies you to intervene in just about any conceivable medical emergency. You've served as a flight surgeon during a previous lunar mission and you have access to dozens of flight surgeons around the world. Your current duties involve providing in-flight health care for three male and three female crewmembers that are onboard the Shackleton, Earth's first electro-magnetic powered[2] spacecraft. The crew is three months into the two-year outbound trip to Callisto. As with all crewmembers assigned to ECMs, the astronauts were subjected to thorough and rigorous medical examinations prior to the flight. So far, everything has been proceeding by the book; you've been monitoring their daily exercise routine, checking their biomedical data, and talking with them on a daily basis. Then, on Mission Day #99, Stewart Hawke, the Shackleton's chief scientist, a 51-year-old Canadian male, requests an unscheduled private medical conference (PMC) with you. The request is received by Mission Control at 4 am on a Sunday morning, so the call is patched through to your house. After a few slugs of coffee, you begin the PMC with Hawke, who is in severe pain. He complains of a sharp stabbing pain in his lower abdominal area, which is now almost unbearable. You conduct the PMC while reviewing his medical history, which reveals that his father suffered from episodes of kidney illness from the age of 45. Hawke also has a sister with a similar history. Hawke has been perfectly healthy throughout the mission and during his previous missions, which included a six-month stay on the surface of the Moon three years previously and a four-week liquid-breathing mission onboard the Atlantica underwater facility in the Challenger Deep. As part of his anti-oxidant regime, he takes a multivitamin together with a biological response modifier as prescribed by mission regulations. He tells you he has not experienced shortness of breath, fatigue, urgency in urinating, vomiting, or diarrhea. The pain is sharp and frequent and usually accompanied by nausea, which prompted Hawke to take two aspirin before making the PMC request. Due to the length of the mission, water is rationed, and Hawke tells you that because of his heavy work schedule, he rarely drinks the prescribed daily two liters of fluid.

Following the PMC, you consider the possible causes of Hawke's symptoms. Potential maladies include appendicitis, but this is immediately ruled out because all ECM crewmembers had their appendices removed before flight as a precaution against just such an event. Inflammation of the stomach is another possibility, as is a small blockage of the bowel. On Earth, a simple diagnostic test could rule out these causes, but medical facilities onboard the Shackleton are limited. On arrival at Mission Control, you request that Petacchi, the Shackleton's CMO, conduct a medical examination of Hawke. You also request that Hawke provide a urine sample. Because of the one-hour time delay in radio transmission, the examination takes almost four hours. At the end of the examination, Hawke is screaming in

---

[2] An electromagnetic thruster, such as the Variable Specific Impulse Magnetoplasma Rocket (VASIMR) being developed by former astronaut, Franklin Chang-Diaz, uses radio waves to ionize and heat a propellant and magnetic fields to accelerate the resulting plasma to generate thrust.

agony and demanding he be given something for the pain. The CMO sends a data packet containing the results of the urine test, which tests positive for blood. You advise Petacchi to break out the Mission Medical Pack (MMP) and instruct him to administer morphine. Then, you inform the Flight Director that Hawke's condition is being evaluated for mission impact and request a flight control team be convened for an emergency medical conference.

The flight control team asks you for your diagnosis based on the PMC. You suggest Hawke may be suffering from a kidney stone infection (Panel 2.2), but until an ultrasound investigation is conducted, it will be impossible to say for sure. You recommend the ultrasound team is brought to Mission Control to conduct the investigation. Petacchi breaks out the mission's ultrasound equipment and performs the examination. The images (Figure 2.2) are transmitted to Mission Control, where they are evaluated by you and a radiologist.

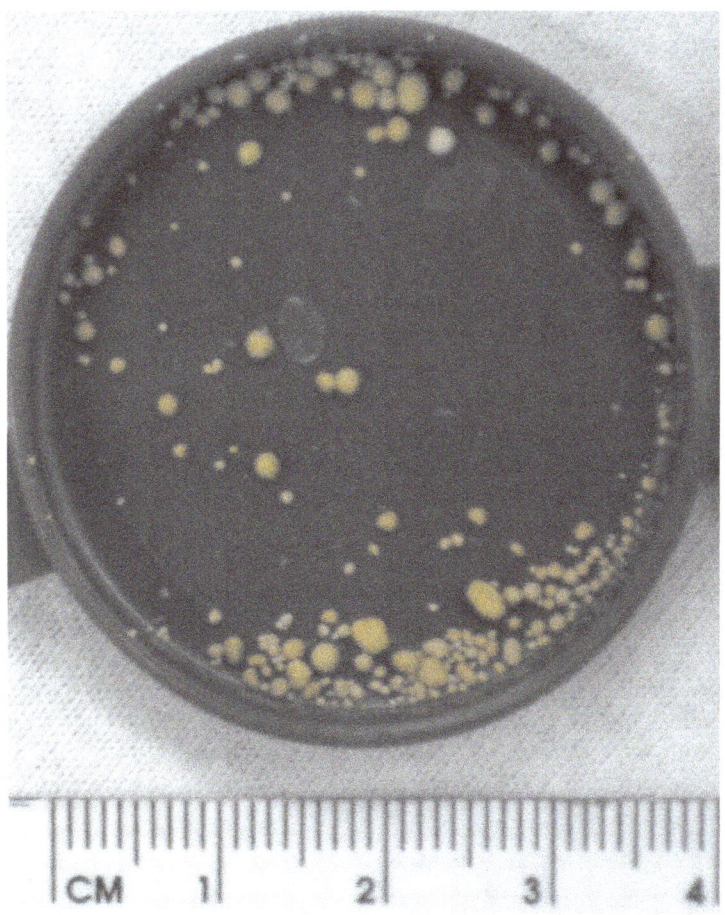

Figure 2.2 Kidney stones. Image courtes

Panel 2.2. Kidney stones

The condition of *kidney stones* results from stones being present in the urethra. Kidney stones can form anywhere within the kidney or bladder and range from tiny microscopic crystals to stones as large as walnuts. They can move from the kidney towards the bladder, causing a number of problems, including excruciating pain. If the stone completely blocks the tube draining the kidney, the kidney could stop functioning. Once renal stones start to move, they can be excruciatingly painful. The stones are solid concretions or calculi (crystal aggregations) formed in the kidneys from dissolved urinary minerals. *Nephrolithiasis* is the medical condition of having kidney stones, while *urolithiasis* refers to the condition of having calculi in the urinary tract (which also includes the kidneys), which may form or pass into the urinary bladder.

Kidney stones often leave the body in the urine stream but if stones grow to sufficient size (two to three millimeters), they may cause obstruction of the urethra. The obstruction causes dilation or stretching of the upper urethra and renal pelvis (the part of the kidney where the urine collects before entering the urethra) as well as muscle spasm of the urethra, trying to move the stone. This leads to pain, most commonly felt in the lower abdomen. There may also be blood in the urine due to damage to the lining of the urinary tract.

Diagnosis can be confirmed by radiological studies or ultrasound examination together with urine and blood tests. The most frequently used procedure for treating kidney stones is extracorporeal shock wave lithotripsy (ESWL), which uses shock waves to break down the stones into small particles, which are passed through the urinary tract in the urine.

For mid and lower-urethra stones, a procedure called *ureteroscopy* may be needed. This involves the surgeon passing a small fiber-optic instrument called a ureteroscope through the urethra and bladder into the urethra. The surgeon locates the stone and either removes it with a cage-like device or shatters it with a special instrument that produces a shock wave.

Astronauts are particularly susceptible to kidney stones because crewmembers lose bone mass during spaceflight and much of the excess calcium ends up in the urine. For example, studies on long-duration crews onboard Mir and the International Space Station (ISS) found astronauts had significantly higher levels of calcium phosphate in urine. The higher calcium levels are probably contributing to the increased calcium-stone-forming potential. They also found fluid intake and urine volume were significantly lower than normal, which means calcium salts are more likely to crystallize and grow into stones.

You examine the ultrasound images and notice a small obstruction in the urethra (the tubes that transport urine from the kidneys to the bladder) – a classic sign of a kidney stone. You send a data packet to Petacchi with instructions on how to commence treatment. Then, you explain the problem to the Flight Director, informing him that it is possible the condition may not be resolved but that treatment has started. Unfortunately, due to the pain, Hawke will be unable to perform any mission duties and will require continuous medical monitoring. If the mission had been onboard the ISS, the default emergency would have been to de-orbit onboard the Dragon capsule. With the crew almost half a billion kilometers away, the options are rather more limited.

Comforted by the fact Hawke has adequate pain control for at least the next 48 hr, you return home in the hope of catching up on some much needed sleep. No sooner are you in the door than your phone rings. It's the Flight Director requesting you return to Mission Control immediately. Hawke's condition has worsened and NASA management has requested an update of the situation because the media are covering the event on television; Fox News has already broadcast an article saying Hawke is just hours from death and the space agency is planning on euthanizing him! You return to Mission Control, where you conduct another PMC with Hawke, who is visibly anxious and deathly pale. Despite the pain killers, Shackleton's chief scientist is complaining that the pain in his stomach is the worst he has ever experienced. You are concerned about the lack of improvement and the deterioration of Hawke's condition. Urine samples are still positive for blood and white cells, suggesting possible urosepsis[3] – a potentially fatal condition. You prescribe narcotics and antibiotics to treat the condition and report to the Flight Director. After you've explained the seriousness of Hawke's condition, the Flight Director asks you for worst and best-case scenarios. Based on the results of the ultrasound and after consulting with Petacchi, you recommend continuing the course of treatment and increasing hydration with intravenous fluids. NASA releases a press statement stating Hawke is expected to make a full recovery and the mission is in no danger. An hour after the press release, CapCom receives word from the Shackleton that Hawke has lapsed into a coma.

You conduct an interview with Petacchi and Everett, Shackleton's commander. Petacchi tells you that Hawke was feverish and disoriented before lapsing into the coma. He is being looked after by the flight engineer and the pilot, which is seriously affecting their duties. You review the latest assessment of Hawke's condition and shake your head. His white blood cell count continues to increase and the course of treatment seems to be having no effect. You present the case to the Multilateral Space Medicine Board (MSMB), who agrees that Hawke probably has less than 48 hr to live. You are about to call the Flight Director to tell him the bad news when you receive a call from Hawke's wife, who has been watching a sensationalist account of her husband's imminent death on Fox News. NASA's public communications officer has told her not to worry, but she wants to hear the truth from you. You tell her that her husband is gravely ill and that everything possible is being done to ensure his

---

[3] Urosepsis is a urinary tract infection/bacterial infection. The bacteria causing the infection ascend the urethra and infect the bladder. Once there, the infection poses the risk of further spreading to the kidneys. The worst consequence is the spreading of the infection to the bloodstream, where it can prove fatal.

survival. Then, you excuse yourself and head into the Flight Director's office. The Flight Director has just received a data package from the Shackleton requesting the crew be given permission to euthanize Hawke in light of his imminent demise. Everett has made a strong case that since Hawke's death is inevitable, valuable life-support resources shouldn't be wasted keeping him alive. The Flight Director shakes his head and tells you that he has informed the Washington DC Representative, Jordan James, Chairman of the House Committee on Science, of the matter. It was James's committee that approved the $40 billion Callisto mission and he's understandably concerned that a crew euthanizing one of its own might turn the American public away from funding future missions. You agree, but you have to side with the commander. You refer the Flight Director to the Interplanetary Bioethics Manual (IBM, see Appendix) and point out the reference to euthanization of critically ill crewmembers. You also remind the Flight Director of the crewmembers' waiving their right to life in the event of just such an eventuality. Given the circumstances, Hawke's death is clearly in the mission's best interests and the Flight Director recommends that the chief scientist's life support be withdrawn. He is over-ruled by Jordan James, but half a billion kilometers away, Shackleton's commander decides to put the integrity of the mission above a Washington bean-counter, and orders Petacchi to withdraw Hawke's life support. Hawke is pronounced dead three hours later. Petacchi's wife is understandably distraught when she is told the news and insists her husband's body be cryogenically preserved so an autopsy can be performed when the Shackleton returns in three years' time. Although the IBM makes provision for such a request, the crew doesn't warm to the prospect of sharing a spacecraft with a corpse for the next three years. The crew request Hawke's body be buried in space. The Flight Director, despite the protestations of Jordan James, agrees. Three hours later, in accordance with the "burial in space" procedures described in the IBM, Everett jettisons Hawke's body into deep space.

Obviously, the death of a crewmember represents the extreme end of the spectrum of the myriad issues facing the flight surgeon during an interplanetary mission. However, it is inevitable that, sooner or later, one or more severe emergency medical events will occur. Some medical events may resolve or at least improve quickly with treatment, others may require continuing care, and yet others may require resuscitative measures. Some will end up like Hawke. The flight surgeon's decision to determine end points will be complex and will be based upon the CMO's best judgment and the guidelines described in the IBM or a similar document. Factors the flight surgeon may consider will include whether there are single or multiple incidents, the resources available to treat the patient, and the operational impact upon the crew of the potential loss or extended disability of the patient. In common with 19th-century polar exploration, ingenuity and determination will be used during the treatment of unusual situations. Other factors the flight surgeon will need to consider include resource utilization in the event of multiple illnesses or casualties, identifying specific roles for caregivers, anticipated end points of treatment, and communication with the ground crew to assist in prioritization of care. Given the unique characteristics of the environment (Figure 2.3), the manner in which the flight surgeon administers these medical intervention strategies will be very different from

Figure 2.3 Crewmembers embarked upon exploration class missions will work in extreme environments that will influence medical intervention strategies. Image courtesy: NASA.

that on Earth, so it is worthwhile taking a look at some of the medical treatment the CMO will be responsible for.

IN-FLIGHT HEALTH CARE

At the time of writing, there has not been a life-threatening medical emergency during a space mission, but there have been several medical events, some of which are listed in Table 2.5. However, ECMs, by virtue of the sheer length of the mission, which may be measured in years, means there is a high probability one or more of the health issues listed in Table 2.6 will require intervention by the CMO. Having said that, much of the CMO's time will be spent dealing with common complaints such as overexertion, strains, and sprains. For example, backaches are a common complaint thought to be associated with elongation in vertebral column length and stress placed on intervertebral discs. To get an idea of what other complaints the flight surgeon may have to face, it is worth looking at what has been learned from other isolated environments on Earth, such as Antarctica and submarine missions; by learning about the type and incidence of medical-surgical and behavioral health events that occur in these environments, it's possible for flight surgeons to plan for the future needs during ECMs. We'll start by discussing submarine[4] missions (Figure 2.4).

---

[4] Submarine missions serve as good analogs for extended-duration space missions. Until April 2005, NASA's total manned spaceflight amounted to 76.57 man-years. By comparison, one Trident Patrol (10 weeks × 7 days/week × 24 hr/day × 155 crewmen = 260,400 man-hours, or 29.7 man-years!

Figure 2.4 Seawolf-class submarine, USS *Connecticut* (SSN-22). Image courtesy: General Dynamics.

Medical events during submarine missions are instructive, as they occur in a confined, remote environment where there is limited diagnostic and therapeutic support. They also occur in an environment where potentially life-threatening events may end a mission. From January 1st, 1997, through December 31st, 1998, the US Navy described the incidence of illnesses and injuries on 136 submarine patrols. The numbers of acute events were related to the total number of person-days under way, with 2,044 acute events in 1.3 million person-days at sea (or 157 acute events per 100,000 person-days). When it came to the more serious events requiring evacuation, a range of 1.9 to 2.3 medical evacuations per 1,000 person-months was reported for all submarines in the US Atlantic Fleet from 1993 to 1996 and a range of 1.8 to 2.6 evacuations per 1,000 person-months was reported for humane reasons (i.e. death in the family).

Another study reviewed health data from 885 Polaris submarine patrols from 1963 to 1973 – a period equivalent to 4,410,000 person-days of submarine activity. During this time, 1,685 medical events resulted in 6,460 duty days lost. The events with the six highest rates of occurrence were, in descending order, trauma, gastrointestinal disease, respiratory infections, dermal disorders, infection, and genitourinary disorders. However, the range of disorders also included all sorts of other medical events, such as arrhythmia, tachycardia, hepatitis, hemorrhage, schizophrenia, appendicitis, and crush injuries. Based on these data, NASA has estimated there may be one major medical event requiring intervention of the type described in the Hawke scenario during a three-year ECM with a crew of six.

Table 2.5. In-flight medical events for US astronauts (STS-1 through STS-89).[5]

Medical event or system by International Classification of Diseases category	Number	Percent
Space adaptation syndrome	788	42.2
Nervous system and sense organs	318	17.0
Digestive system	163	8.7
Skin and subcutaneous tissue	151	8.1
Injuries or trauma	141	7.6
Musculoskeletal system and connective tissue	132	7.1
Respiratory system	83	4.4
Behavioral signs and symptoms	34	1.8
Infectious diseases	26	1.4
Genitourinary system	23	1.2

Table 2.6. Major health and medical issues during exploration class missions.

Health issue	Risk	Health issue	Risk
Radiation protection	Severe	Neurovestibular	Severe
Hearing conservation	Moderate	Habitability	Moderate
Cardiovascular	Moderate	Extravehicular activity	Severe
Muscle loss	Moderate	Psychological	Moderate
Bone loss	Severe	Behavioral	Moderate

NASA's prediction has been backed up by similar data compiled from experience in the Antarctic. The Australian National Antarctic Research Expeditions (ANARE) Health Register compiled 1,967 person-years of data from 1988 to 1997. It documented 5,103 illnesses and 3,910 injuries, the distribution and variety of which were similar to spaceflight data. The Health Register also noted several deaths resulting from drowning and exposure, appendicitis, cerebral hemorrhage, acute myocardial infarction, carbon monoxide poisoning, and burns. While it is unlikely an astronaut will drown, each of the other circumstances is a potential medical event on a spacecraft, indicating the wide variety of medical emergencies that a flight surgeon will need to consider when planning for the health care management techniques described in the next section.

Health care management en route to a distant planet will follow many of the procedures here on Earth. For example, the starting point for medical care will be a description of the chief complaint and a physical examination. However, unlike when you visit the doctor here on Earth, the physical examination technique

---

[5] Adapted from Billica, R. In-Flight Medical Events for US Astronauts during Space Shuttle Program STS-1 through STS-89. April 1981–January 1998. Presentation to the Institute of Medicine Committee on Creating a Vision for Space Medicine During Travel Beyond Earth Orbit, February 22, Johnson Space Center, Houston (2000).

onboard a spacecraft will be adapted to the microgravity environment. That's because the methods of locating internal organs and other diagnostic methods are affected by the lack of gravity, which means the astronaut, the CMO, and the equipment have to be stabilized.

Within the spectrum of providing health care to the astronauts, the CMO will also be responsible for other factors that may affect the well-being of the crew. For example, he/she will be responsible for ensuring the crew's dietary needs are met and checking crewmembers are drinking enough fluid to prevent the formation of kidney stones, thereby avoiding a situation like Hawke's. He/she will also need to be prepared to deal with the consequences of accidental leaks of chemical elements into the spacecraft atmosphere. Experience in LEO has shown that environmental hazards come from several sources, the dominant source being heat degradation of electronic devices, which produces formaldehyde and ammonia. Needless to say, exposure to these sorts of chemicals could result in multiple casualties that would soon outstrip the finite resources available on the spacecraft, so the CMO will need to be able to identify acute signs and symptoms early. Another monitoring task for the CMO will be checking how much radiation each crewmember has absorbed. This will be particularly important because while much is known about the radiation environment of LEO, little is known about the radiation environment of interplanetary space, where quantities of solar and galactic radiation and the potential for exposure increase. So, when the CMO isn't performing routine check-ups and reminding the crew to drink, they will turn their attention to other daily tasks such as checking how much radiation each crewmember has absorbed, by checking the crew's dosimeters. To protect the crew against excessive radiation exposure, the CMO will no doubt periodically check with the Space Radiation Analysis Group (SRAG)[6] back in Houston.

Other routine check-ups will include making sure astronauts keep up their exercise regime and monitoring crewmembers' cardiovascular integrity, which will mean checking for a whole range of cardiovascular symptoms and abnormalities. For example, astronauts during an ECM may suffer high blood pressure, ventricular premature beats, atrial arrhythmias, tachycardia, chest pain, shortness of breath, syncope, and a host of other heart-related symptoms. Once the CMO has finished those checks, they may turn their attention to the crew's dental health. Although astronauts will be pre-screened, even good teeth and a history of preventive care can't guarantee that no caries will develop in anyone over the course of a multi-year mission. As with most space missions, astronauts will be subject to a heavy work schedule, which means, occasionally, they may not always maintain good dental hygiene (Panel 2.3). This, combined with the lack of foods with natural gingival-cleansing properties, means the CMO will probably have to fill the odd cavity (ANARE data reported dental events accounted for 8.80% of all medical events despite crewmembers having been pre-screened and found to be dentally fit).

[6] SRAG team members examine space weather data, reports, and forecasts for trends or conditions that may produce enhancements to the near-Earth radiation environment; they then report the information to flight management.

> **Panel 2.3.** Toothache in orbit
>
> In 1978, Soviet cosmonaut Yuri Romanenko experienced a toothache during the 96-day flight of *Salyut 6*. To begin with, Romanenko took overdoses of pain killers to deaden a toothache that was causing his eyes to literally roll with pain. Because he considered it would be a disgrace to complain, Romanenko didn't report his discomfort to the ground for two weeks. As his problem worsened, his fellow crewmembers pleaded for help from the ground but the Soviet space program had no contingency plans for dental emergencies. In fact, the only advice from controllers was for Romanenko to take a mouthwash and keep warm! Romanenko suffered for two more weeks before *Salyut 6* touched down on schedule. His ordeal was subsequently the subject of a televised interview in the Soviet Union and featured in published accounts in Russian and US space and dental literature.

Gastrointestinal problems are another common ailment among astronauts, accounting for up to 8% of the recorded medical events on Space Shuttle missions. Many of the problems are caused by the microgravity environment, which causes some astronauts to be constipated and others to suffer from diarrhea. While the CMO will treat these with common over-the-counter medications such as Imodium and Pepto Bismol, other gastrointestinal problems will require more aggressive courses of treatment and, in some cases, even surgery. For example, an obstruction of the gallbladder or appendix that becomes infected can be lethal without operative intervention. Given the seriousness of such an event, astronauts will probably have their appendices removed before flight (an "elective appendectomy" in medical parlance), so this shouldn't be a problem, but other problems such as an inflammation of the pancreas (a life-threatening condition even with the best medical care) may cause the CMO some sleepless nights.

Yet another regular feature of the CMO's job will be handing out pills. Lots of pills. Many of these pills will simply be countermeasures against all the adaptations to microgravity. For example, space missions result in a decreased red blood cell (RBC) mass, which means astronauts become anemic. To combat this, astronauts will probably be given erythropoietin, a hormone that promotes RBC survival. To boost the astronaut's immune systems, the CMO will also hand out immune system boosters such as Eleutherococcus Senticosus (ES), a supplement used by the Russian space program since 1966. ES not only boosts the immune system, but also helps the body adapt to and cope with unfavorable conditions, such as physical and psychological stress, infections, environmental pollutants, radiation, and extreme climatic conditions.

In addition to acting as the mission's dentist, pharmacist, and doctor, the CMO will also be expected to serve as a behavioral health specialist. Astronauts embarked upon multi-year missions will be exposed to the most isolated, hostile, and confined

32 Interplanetary health care

environment in human exploration history and it is inevitable that sooner or later, cracks will appear in even the toughest crewmembers. Some may experience depression, others may become overly anxious, while some may become psychotic. Some of these mental health issues may simply be caused by the cumulative effects of environmental and interpersonal stressors that become magnified by the sheer length of the mission. In either case, it will be up to the CMO to intervene. Most likely, he/she will take advantage of psychiatric expertise on the ground and, if necessary, make use of psychotropic medications (or a straitjacket) onboard.

ANESTHESIA

Inevitably, during ECMs, anesthesia and pain management will be required for unanticipated accidents and various medical conditions. For the CMO, this aspect of health care will present major challenges. For example, little is known about gas diffusion in reduced gravity – a lack of knowledge that may compromise the administration of anesthesia and pain management procedure. Also, in microgravity, fluids and gases do not separate on the basis of differing densities. Consequently, a vial of a drug or a bag of intravenous fluid looks something like shaving foam! Furthermore, many devices that depend on gravity-induced separation of gases and fluids, such as anesthetic vaporizers, simply break down in microgravity. To overcome these challenges, future CMOs will need to adapt current anesthesia techniques and procedures to meet the unique problems that arise when administering anesthesia in space to microgravity-exposed patients.

AIRWAY MANAGEMENT

Another technique CMOs will need to be proficient in administering is airway management. However, success in applying some of the most effective methods of airway management requires frequent and regular practice, especially in the microgravity environment (Figure 2.5).

The types of airway equipment and techniques required will be based on the type of surgery and trauma anticipated. Before such equipment and techniques are selected, it will be necessary to decide which long-term airway care procedures might be needed. For example, to give injured astronauts the best chance of survival, CMOs will almost certainly decide to use anesthetic techniques that do not require endotracheal intubation. You may have seen this procedure in the movies. It refers to the placement of a flexible plastic tube into the trachea to protect the patient's airway and provide a means of mechanical ventilation. Even on Earth, it is a procedure that is potentially dangerous and requires a great deal of clinical experience to master: if performed improperly (in microgravity studies onboard parabolic flights, the procedure was unsuccessful in 15% of situations), complications may lead to the patient's death. Given the potential dangers of such a procedure, flight surgeons will probably recommend that CMOs use a laryngeal

Figure 2.5 This zero-G intubation was performed during parabolic flights conducted on an Airbus 300 over the Atlantic Ocean. The patient's head was gripped between the anesthesiologist's knees, with the torso strapped to the surface. Three personnel with no experience in airway management or microgravity participated in the study, which attempted the procedure seven times. Image courtesy: European Space Agency.

mask, the use of which is technically much easier. The mask also provides an excellent airway and in the simulated atmosphere of microgravity, its use has been more successful.

SURGERY

Surgery also presents interplanetary CMOs with a set of unknowns. For example, the physiological changes that occur in astronauts embarked upon long-duration space missions may affect wound healing and resistance to infection, each of which is crucial for recovery from surgery. Another uncertainty is the response to hemorrhage and fluid resuscitation, which will probably be altered by the effects of microgravity. Then, there's the problem of bleeding in zero gravity. Research onboard parabolic aircraft has shown that bleeding increases during surgery. On Earth, gravitational forces help collapse the veins and help stop the flow of blood, but in microgravity, the gravitational force is absent and external compression must be supplied. No doubt, the CMO will adapt and employ a certain level of ingenuity to overcome these challenges, but what happens if the CMO has to perform a surgical procedure for which there are no instruments? Remember, therapeutic options will be limited by the medical equipment carried onboard and space onboard an interplanetary spacecraft will be at a premium. Fortunately, thanks to computer-assisted design and computer-assisted manipulation (CAD–CAM) technology, it will be possible to fabricate tools to order from specifications contained in an onboard database or transmitted from Earth. So, a CMO requiring a seldom-used surgical instrument a billion kilometers from home could simply custom order the tool using the onboard CAD–CAM equipment. A similar system (Panel 2.4) will be used by CMOs who need to replace the organs or limbs of an injured astronaut.

So far, we've discussed the problems of stopping bleeding, the unavailability of instruments, and wound healing, but what about performing actual surgery? In the microgravity environment of an interplanetary spacecraft or the reduced gravity of a

---

Panel 2.4. Tissue-engineered Organ Replacement System (TORS)

Tissue engineering has long held promise for building new organs to replace damaged livers, blood vessels, and other body parts. The TORS encapsulates living cells in cubes and arranges them into 3-D structures, just as a child constructs buildings out of Lego blocks. The technique, dubbed "micro-masonry", employs a gel-like material that acts like concrete, binding the cell "bricks" together as it hardens. The tiny cell bricks allow scientists to build artificial tissue such as organs and limbs. To obtain single cells for tissue engineering, researchers first break tissue apart, using enzymes that digest the extracellular material that normally holds cells together. Some scientists have successfully built simple tissues such as skin, cartilage, or bladder on biodegradable foam scaffolds, although the tissues don't have the same complexity as normal tissues. The tissues are built "biological Legos" by encapsulating cells within a polymer called polyethylene glycol (PEG), which is a liquid that becomes a gel when illuminated. So, when the PEG-coated cells are exposed to light, the polymer hardens and encases the cells in cubes. Once the cells are in cube form, they can be arranged in specific shapes using templates made of a silicon-based polymer. Both template and cell cubes are coated again with the PEG polymer, which acts as a glue that holds the cubes together as they pack themselves tightly onto the scaffold surface. Once the cubes are arranged properly, they are illuminated again, and the liquid holding the cubes together solidifies.

It may sound very "Brave New World", but a basic TORS is already on the drawing board and on future HOPE-type ECMs, such an organ printing system may help astronauts grow everything from an artificial liver to a new pancreas.

Jovian moon, problems such as anchoring both the patient (just look at Figure 2.5) and the operating team, maintaining a sterile field, and controlling blood and body fluids can be expected. To minimize these problems, CMOs will opt for minimally invasive procedures such as laparoscopic surgery. Laparoscopic surgery is a modern surgical technique in which operations are performed through small incisions (usually 0.5–1.5 cm) within the abdominal or pelvic cavities. Because the incisions are so small, recovery is quick and body fluids are contained more effectively than in traditional surgery. The procedure has been tested successfully on pigs during parabolic flight.

While laparoscopic surgery may be an effective procedure for many types of surgery, there are some injuries that may exceed the medical capabilities of the spacecraft and the capacities of the CMO to intervene medically. For example, closed head injuries and spinal cord injuries may represent severe life-threatening events, since management and treatment of astronauts with these types of injuries would likely be beyond the capability of even the most experienced CMO. Individuals with mild or moderate closed head injuries may survive but remain disabled because of residual neurological deficits. Terrestrial management issues today include placement of burr holes for evacuation of subdural hematomas, feeding and airway control, spinal cord stabilization, and management of bowel and bladder functions and infections, none of which will be available in the cramped confines of an interplanetary spacecraft or surface habitat.

REHABILITATING ASTRONAUTS

When the astronauts finally make it to the surface of Callisto or Mars and eventually back to Earth – hopefully without any medical emergencies – they will begin a post-flight rehabilitation program. The program will be similar to the plan described in Table 2.7. As you can see, the CMO will play a prominent role in ensuring deconditioned astronauts restore their preflight muscle strength and aerobic capacity.

Table 2.7. Post-flight rehabilitation plan.

Description	The post-flight rehabilitation plan is a three-phase plan designed to protect the health and safety of astronauts following landing and on returning from interplanetary missions and to actively assist in the crewmembers' return to preflight health and fitness levels		
Schedule		*Outbound post landing*	
	Duration	*Schedule*	*Personnel*
	Rehabilitation Phase I	0–3 days post landing	CMO, crewmembers, and crew surgeon via groundlink
	120 min/day		
	Proprioceptive neuromuscular facilitation (PNF) techniques, massage, and light manual resistance exercises		

Rehabilitation Phase II 120–150 min/day Agility and coordination tasks, light cardiovascular exercise. Massage, PNF techniques, flexibility, and strength exercise	4–10 days post landing	CMO, crewmembers, and crew surgeon via groundlink
Rehabilitation Phase III 150–180 min/day Agility and coordination tasks. Cardiovascular exercise. PNF techniques, massage, and strength exercises	11–14 days post landing	CMO, crewmembers, and crew surgeon via groundlink
Rehabilitation Phase IV 90–120 min/day Cardiovascular and strength training. Massage. Fitness testing once per week	15–21 days post landing	CMO, crewmembers, and crew surgeon via groundlink
Inbound post landing		
Rehabilitation Phase I 120 min/day Assisted walking. Hydrotherapy, proprioceptive neuromuscular facilitation (PNF) techniques, massage, and light manual resistance exercises	0–7 days post landing	Astronaut Strength Conditioning & Rehabilitation Staff (ASCR), crewmembers, and crew surgeon
Rehabilitation Phase II 120–150 min/day Assisted walking. Hydrotherapy, agility, and coordination tasks, light cardiovascular exercise. Massage, PNF techniques, flexibility and strength exercise	8–30 days post landing	ASCR, crewmembers, and crew surgeon
Rehabilitation Phase III 150–180 min/day Agility and coordination tasks. Cardiovascular exercise. PNF techniques, massage, hydrotherapy and strength exercises	31–60 days post landing	ASCR, crewmembers, and crew surgeon
Rehabilitation Phase IV 90–120 min/day Cardiovascular and strength training. Massage. Fitness testing once per week	61–120 days post landing	ASCR, crewmember, and crew surgeon

Special requirements	Crewmembers will perform rehabilitation on duty days only (5 day/week)
	Medical status checks will be performed once per week
	The ASCR and Exercise Physiology Laboratory will make recommendations to the crew surgeon regarding rehabilitation progress and exercise certification of crewmembers
Notes	During each rehabilitation phase, crewmembers will be assessed using fitness tests to evaluate isokinetic function, oxygen uptake, agility, and coordination and flexibility

Astronaut health during ECMs will require a continuum of preventive, therapeutic, and rehabilitative care on the ground, during the mission, and upon return to Earth. The continuum will include normal health maintenance and care for the physiological adaptations astronauts will be exposed to as a result of the extreme environment of space. Each of these phases will require the skills of an experienced CMO and ground-based flight surgeons capable of responding to the myriad minor and major medical problems that can develop among members of a group of individuals over extended periods of time in an extreme environment.

REFERENCES

[1] Dehart, R.L.; Davis, J.R. (eds). *Fundamentals of Aerospace Medicine*, 3rd edn, pp. 617–618. Lippincott Williams and Williams, Philadelphia (2002).

3

Medical qualification for exploration class missions

Exploration class missions (ECMs) will demand unique and extreme medical selection criteria. Obviously, with so many stressors, it will be more important than ever to ensure that medical selection criteria identify only the most hardy and resilient crewmembers. To do this, space agencies will first implement screening methods to minimize risk factors for diseases that cannot be treated during such ventures. While these methods are currently undefined, it can be expected that future ECM astronauts will be subject to an assessment of genetic factors to screen for radiation resistance and bone density. The next step the space agencies will most likely take will be to demand that the elite cadre of interplanetary crewmembers undergo precautionary surgery to remove (among other things) their appendices. And, of course, given the long time these crews will be expected to share a cabin no larger than a school bus for the best part of three years or more, it is likely a revised psychiatric assessment will also be part of the selection.

While astronauts in low Earth orbit (LEO) have coped well with the typical problems of living in space, the physical demands, dangers, and discomforts of spaceflight multiply as mission length increases. Astronauts bound for destinations such as Mars, Saturn, and beyond will need to deal with an increased risk of cosmic and solar radiation, shrinking muscles, brittle bones, and social tensions caused by months of confinement and isolation. There will be precious few amenities, no fresh fruit or vegetables, and limited forms of entertainment. Taken in small doses, none of these issues seems likely to cause significant problems. But add them all up and multiply them by several months or years in deep space and it becomes easy to understand how even the best adjusted, even-tempered astronaut might behave unpredictably once in a while.

MEDICAL SELECTION OF ASTRONAUTS

The purpose of medical standards (which will probably be similar to the ones listed in Table 3.1) for ECMs will be to ensure astronauts are physically and temperamentally fit for the performance of orbital activities and extended operations on a planetary surface. As we've already seen in the first two chapters of this book, spending long periods of time in deep space will be extremely stressful, so any defects and diseases will automatically disqualify potential crewmembers.

Table 3.1. Exploration class medical examinations and parameters.

Item	Description
1	Medical history
	a. Medical survey
	b. Genetic history/DNA test
	(i) Screened for future disease(s)
	(ii) Screened for resistance to radiation
	(iii) Screened for high bone density
	c. Questionnaire
2	Physical examination
	a. General physical
	b. Anthropometry (biometric assessment of the body)
	c. Muscle mass
	d. Pelvic exam and Pap smear
	e. Procto sigmoidoscopy (invasive examination of the large intestine from the rectum through the last part of the colon)
3	Cardiopulmonary evaluation
	a. History and examination
	b. Physical fitness test
	c. Exercise stress test
	d. Blood pressure
	e. Resting and 24-hr electrocardiograph (ECG)
	f. Echocardiogram (ultrasound technique used to generate 3-D image of the heart)
4	Ear, nose and throat (ENT) evaluation
	a. History and examination
	b. Audiometry
	c. Tympanometry (an objective test of middle-ear function; it is not a hearing test, but a measure of energy transmission through the middle ear)
5	Ophthalmological evaluation
	a. Visual acuity, refraction, and accommodation
	b. Colour and depth perception
	c. Phorias (the relative directions of the eyes during binocular fixation on a given object in the absence of an adequate stimulus)
	d. Tonometry (procedure ophthalmologists perform to determine the intraocular pressure (IOP) – the fluid pressure inside the eye)
	e. Perimetry (systematic measurement of differential light sensitivity in the visual field by the detection of the presence of test targets on a defined background) and retinal photograph
	f. Endoscopy
6	Dental examination
	a. Panorex (also known as an "orthopantogram", a panorex is a panoramic scanning dental X-ray of the upper and lower jaw showing a 2-D view of a half-circle from ear to ear) and full dental X-rays within last two years
7	Neurological examination

a. History and examination
b. EEG at rest
8. Psychiatric/psychological evaluation
   a. Psychiatric interviews
   b. Psychological tests
9. Radiographic evaluation
   a. Chest X-ray
   b. X-ray DNS
   c. Mammography
   d. Medical radiation exposure history and interview
   e. Abdominal and urogenital ultrasonography (an ultrasound-based diagnostic imaging technique used to visualize subcutaneous body structures)
10. Laboratory investigation
    a. Complete hemogram (blood test, including an estimate of the blood hemoglobin level, packed cell volume, and blood count)
    b. Blood biochemistry
    c. Immunology
    d. Serology (scientific study of blood serum; the term usually refers to the diagnostic identification of antibodies in the serum)
    e. Endocrinology
    f. Urinanalysis
    g. 24-hr chemistry
    h. Renal stone profile
    i. Urine endocrinology
    j. Urine reticulocytes (RE)
    k. Stool reticulocytes (RE)
    l. Occult blood
    m. Ova and parasites
11. Other tests and parameters
    a. Drug screen
    b. Montoux test (test for tuberculosis)
    c. Microbiological, fungal, and viral tests
    d. Pregnancy test
    e. Screening for STD
    f. Abdominal ultrasonography
    g. Sterilization and sperm/egg banking
12. Pre-emptive surgery requirements
    a. Appendectomy (surgical removal of the appendix)
    b. Cholecystectomy (surgical removal of the gallbladder)

For those hoping to be selected for an interplanetary mission, the first step will be a medical assessment, which will include an evaluation of each of the following physiological systems and conditions:

1. endocrine system: this is the system of glands that releases hormones that affect almost every cell, organ, and function in the body;
2. genitourinary system: this organ system includes the reproductive organs and the urinary system;

3. respiratory system: includes the airways, lungs, and the respiratory muscles;
4. cardiovascular system: comprises the heart and the blood vessels;
5. gastrointestinal system: this system is divided into upper and lower parts; the upper gastrointestinal tract consists of the esophagus, stomach, and duodenum while the lower gastrointestinal tract includes most of the small intestine and all of the large intestine;
6. neurological: the nervous system is divided into two systems – the central nervous system (CNS) and the peripheral nervous system (PNS); the spinal cord and the brain make up the CNS while the peripheral nervous system is made up of all of the nerves;
7. psychological and psychiatric evaluation: this involves a series of interviews and tests designed to select out those applicants who possess qualities that might represent a risk for behavioral health in space;
8. ophthalmology: this is the branch of medicine that deals with the anatomy, functions, and diseases of the eye;
9. ear, nose, throat, and equilibrium: ear, nose, and throat (usually abbreviated to ENT) disorders are diagnosed by otolaryngologists – a term that derives from otolaryngology, the particular branch of medicine that specializes in ENT;
10. musculoskeletal: this system comprises the body's bones (the skeleton), muscles, cartilage, tendons, ligaments, joints, and other connective tissue (this is the tissue that supports and binds tissues and organs together);
11. hematological and immunologic: hematology is a branch of internal medicine concerned with the study of blood, the blood-forming organs, and blood diseases, while the immune system is designed to defend the body against bacteria, microbes, viruses, toxins, and parasites;
12. general medical condition.

For each of the physiological systems listed, potential astronauts will be required to be free from any system-specific disorder that accredited medical conclusion indicates would render the crewmember unable to perform the duties required during a multi-year mission. For each system, there are a number of disqualifying conditions, some of which are discussed here. We'll start with the endocrine system. The most common disqualifying condition associated with this system is Type 1 Diabetes Mellitus, since an individual with this condition requires injections of exogenous insulin to properly metabolize carbohydrates and lipids. Such a situation is clearly incompatible with extended duration spaceflight since, in the absence of treatment, a potential exists for disastrous incapacitation, which might jeopardize the individual and the crew.

Next, we'll take a look at the respiratory system, for which there are a number of disqualifying conditions. For example, any pulmonary function disorder increases an astronaut's chances of being incapacitated. Imagine an individual who suffers from chronic bronchitis or emphysema; not only would this individual experience significant hypoxia, a situation aggravated by the hypoxic environment of the spacecraft, but in the event of a rapid/explosive decompression, their weakened lung tissue would be at greater risk of damage due to rapid and/or excessive pressure changes.

Individuals with a faulty cardiovascular system will also be disqualified, since the majority of cardiovascular disorders are associated with the risk of sudden death or incapacitation. For example, coronary artery disease (CAD) is unpredictable and may be aggravated by circumstances such as heat, hypoxia, and exposure to high gravitational forces, each of which increases oxygen demand. Another serious cardiovascular disorder is myocardial infarction – a condition associated with atheromatous plaques that have the potential to rupture and occlude vessels; such a situation obviously has the risk of a potentially catastrophic and incapacitating event.

The gastrointestinal system also has to be working well because there a number of conditions that have implications for spaceflight. For example, volume changes of gases inside the stomach during a rapid decompression might aggravate an existing condition and incapacitate a crewmember to such a degree that the astronaut's ability to perform an emergency egress is compromised.

Psychological health will be subject to particular scrutiny. Anxiety disorders, mood disorders, and undesirable personality traits are some of the most common reasons for disqualification during the selection process of European Space Agency (ESA), Canadian Space Agency (CSA), and NASA astronauts. Any of these disorders are incompatible with spaceflight – extended or short-duration – due to their potential to manifest themselves in overt acts such as panic and phobias. Also, these disorders are unpredictable and may be triggered by all sorts of stressful events that have the potential to be disabling. If you've watched science fiction movies like *Sunshine* or *Moon*, you could be forgiven for thinking that deep space will cause all sorts of psychological problems. Well, extended-duration flights (defined as flights longer than 100 days – about 1/10th the anticipated duration of a Mars mission) do expose some problems, but they're not as serious as those portrayed in Hollywood movies. Astronauts will suffer from boredom, fatigue, and circadian rhythm and sleep disturbances, but none of these conditions has the potential to adversely affect the mission. Unfortunately, this hasn't stopped psychologists from arguing that these problems, coupled with the exacting human performance requirements of such missions, constitute risk factors for the development of all sorts of syndromes. They argue that on missions beyond Earth orbit, in which spacecraft crews will be isolated and confined to a small living area and in which medical evacuation will not be an option, the development of these and other mental health problems may exert cumulative detrimental effects on astronauts and on their fellow crewmembers sufficient to jeopardize the mission. In reality, nothing could be further from the truth.

While many (too many!) psychologists argue that little is known about the psychological capacity of humans to withstand the stresses of long-duration space travel, in reality, the opposite is true. You see, such a line of reasoning completely ignores a wealth of experience gleaned from analog environments such as polar exploration. A century ago, in the Golden Age of Exploration, explorers such as Fridtjof Nansen, Ernest Shackleton (Figure 3.1), Roald Amundsen, and Douglas Mawson were household names, boldly going where no man had gone before. They embarked on multi-year missions to the Arctic and Antarctic, and did so without the use of cell phones, global positioning systems, MP3 players, satellite navigation, or psychological assessments. And guess what? The missions went very well for the

Figure 3.1 Sir Ernest Shackleton. Image courtesy: National Oceanic and Atmospheric Administration.

most part, especially considering the long winter nights that lasted several months and a wind chill that was regularly measured in the triple digits. Sure, there were a few arguments and spats, but nobody had a meltdown. And don't forget, most of the crewmembers had been recruited straight off the dock. They didn't have the benefit of a university education or an army of behavioral psychiatrists. So, when humans finally do embark on an interplanetary mission, psychological problems should be the very least of a mission planner's worries, although that's not to say the assessment should be ignored. Nevertheless, a psychiatric selection strategy (Panel 3.1) will be needed to remove unsuitable candidates – an approach that has long been recognized as instrumental for selecting candidates likely to perform and adapt optimally to space.

Panel 3.1. Psychiatric assessment

Psychiatric assessment identifies certain "select-out" disqualifying medical criteria such as schizophrenia and psychopathic tendencies. To determine whether an individual has a history of any disqualifying disorders, a formal clinical evaluation in the form of a psychiatric interview is performed. Other evaluation tools that may be applied include psychometric tests such as the Minnesota Multiphasic Personality Inventory (MMPI), the Rorschach Ink Blot test, and the Million Clinical Multiphasic Inventory (MCMI). Psychiatric interviews are normally conducted with at least two independent psychiatrists. The interviews are carefully structured to ensure candidates provide as much clinical information as possible regarding a particular subject and are presented in a format designed to reduce the number of "no" and "yes" responses. For example, rather than simply asking "Have you ever been depressed?", to which most candidates will respond in the negative, the question is phrased "Tell me about the time when you have been the most sad in your life", thereby ensuring the candidate provides clinical information regarding the subject of depression.

Also, given the recent media attention in 2007 devoted to the bizarre case of astronaut, Lisa Nowak (Panel 3.2 and Figure 3.2), it is likely that NASA will devote particular attention to psychological screening when selecting those who venture beyond LEO.

Panel 3.2. Lisa Nowak

It took NASA astronaut, Lisa Nowak, 12 days, 18 hr, 37 min and 54 sec to secure her place in one of the world's most elite clubs when she flew aboard the Space Shuttle *Discovery* during mission STS-121 in July 2006. It took her about 14 hr to destroy it. That was how long it took the 43-year-old mission specialist to drive the 1,500 km from Houston, Texas, to Orlando, Florida, carrying with her a carbon-dioxide-powered pellet gun, a folding knife, pepper spray, a steel mallet, and $600 in cash. Nowak had discovered that Colleen Shipman, a US Air Force captain, was flying in from Houston to Orlando that night and Nowak wanted to be there to "scare her" into talking about her relationship with the man at the center of a love triangle. That man was Bill Oefelein, who underwent astronaut training with Nowak and, like her, went into space for the first time in 2006, although they had never flown together.

Shipman allegedly saw Nowak, whom she had never met before, wearing a trench-coat, dark glasses, and the wig, following her on a bus from an airport lounge to a car park. Afraid, she hurried to her car. She could hear running footsteps behind her and as she slammed the door, Nowak slapped the window and tried to pull the door open. "Can you help me, please? My boyfriend was supposed to pick me up and he is not here," Nowak was alleged to have pleaded. When Shipman said she couldn't help, the astronaut started to cry. Shipman wound down her window, at which point Nowak discharged the pepper spray. Shipman drove off, her eyes burning, and raised the alarm. Nowak was subsequently charged with attempted first-degree murder in what quickly became the most bizarre incident involving any of NASA's active-duty astronauts.

To say the group to which Nowak belonged (her assignment to the space agency was terminated by NASA on March 8th, 2007) is select is an understatement. Up to 2007, NASA had selected just 321 astronauts since the US agency began preparing to go into space in 1959. She had been subjected to NASA's rigorous screening process and trained for 10 years to cope with the intense stress of spaceflight before her flight. Like all the other astronauts, Nowak had been subject to extensive psychiatric and psychological screening, all of which made her behavior incomprehensible.

To many, the Nowak scandal called to mind every bad science fiction movie in which they send unstable characters into space. Others argued that NASA should have noticed the signs of Nowak's unraveling. These people might have had a point, but you have to remember that people in highly stressful jobs are generally over-achievers who put a high value on performance and a low value on self-care beyond that required to perform the job. These types – astronauts – do a great job ignoring and denying signs of fatigue, either physical or psychological. Instead, they assume a machine-like thought process to deal with any problems. But the human brain isn't just a thinking machine; it is also the seat of emotions, and the suppression of emotions plays out in the battlefield of the subconscious mind. That suppression and the associated physical and psychological damage eventually surface in skewed thought processes and actions, which is exactly what happened to Nowak. From our perspective, Nowak's actions appeared crazy, but her perception of her actions appeared to be a logical way to resolve her problem.

Captain Nowak's drama played out in an airport parking lot. Imagine a comparable scene at a base on the Moon or on a spaceship en route to Mars.

Figure 3.2 Lisa Nowak. Image courtesy: NASA.

Okay, enough about psychiatry and psychology. Let's move on to the next system: ophthalmology. No question, excellent vision will be a requirement for planetary-bound astronauts. Deep space is a hostile environment, characterized by zero gravity, hypoxia, possible ebullism, high-speed acceleration, and electromagnetic glare, each of which has the potential to degrade vision and, in individuals with certain disorders, the potential to blind and disable. Each crewmember will need to perform tasks requiring adequate depth perception, color vision, and spatial discrimination. A deficiency in any one of these functions may jeopardize other crewmembers in the event of an emergency and, on the occurrence of an emergency decompression, those predisposed to visual defects may suffer transitory hemiplegia and, in rare cases, permanent visual impairment.

It goes without saying that hearing, balance, speech and communication, and unrestricted breathing will be especially important. Certain disorders, such as vertigo, for example, may present significant risk to other crewmembers, as the condition may be associated with intractable symptoms that include episodic vertigo and fluctuating hearing loss. Another serious disorder is benign positional vertigo – a condition occurring without warning and associated with vertigo, nausea, and vomiting. Needless to say, an individual with this condition, which would be greatly aggravated in microgravity, would present a serious risk to other crewmembers.

Once the potential crewmember has had all their physiological systems checked, they can move on to the next stage: genetic screening.

GENETIC SCREENING

You may be wondering why space agencies will need to implement genetic screening. After all, aren't the medical standards for astronauts already among the most rigorous in the world? Unfortunately, despite the very high medical standards, current assessments can't predict which crewmembers will fare well when exposed to radiation or which astronauts will recover quickly after losing bone mass. And, as we've seen in the first chapter, the answers to these questions are important because two of the most serious dangers faced by interplanetary astronauts are radiation exposure and bone loss.

Crewmembers embarked upon long-duration missions may approach or exceed career radiation limits due to the harsh radiation environment. Because of the risks of excessive radiation exposure, it just makes sense for space agencies to select crews who are less susceptible to exposure to radiation. Similarly, some people lose bone at faster or slower rates than others.[1] From what we know during long missions onboard the International Space Station (ISS), astronauts lose bone at a rate of up to 1–2% per month in load-bearing bones – a rate six times greater than the rate of

---

[1] At one extreme, NASA astronaut David Wolf lost 12% of his bone mass during his four-and-a-half-month stint on the ISS, whereas some long-duration cosmonauts have not shown any significant bone loss.

48 Medical qualification for exploration class missions

bone loss of women with osteoporosis! By the time the astronauts arrive on Mars, the rate of bone degradation would be so devastating that crewmembers might suffer severe fractures within minutes of stepping on the surface! Studies have shown that inherited factors account for up to 80% of bone mass density variability, so it makes sense to use genetic screening to select those crewmembers with high bone density.

Current international legislation bars employers from using the genetic information of individuals when making hiring decisions. However, space agencies tasked with the task of selecting perhaps the most unique space crew in history will surely be exempted from this requirement. Also, given the unusual nature of interplanetary medical criteria, those selected will cross a legally defined boundary; in the same way as a soldier relinquishes certain individual rights when joining the military, astronauts chosen for a multi-year mission will be expected to do the same and accept collective standards contributing to the common good of realizing a successful mission.

In addition to screening for radiation susceptibility and bone strength, prospective ECM astronauts, just like Ethan Hawke's character in *Gattaca* (Figure 3.3), will also be genetically tested to diagnose vulnerabilities to inherited diseases that may cause problems during long missions. The testing will also reveal information concerning the presence of genetic diseases and mutant forms of genes associated with increased risk of developing genetic disorders. Additionally, genetic testing will confirm or deny a suspected genetic condition and provide information concerning the possibility that an astronaut may develop a disorder.

Figure 3.3 Ethan Hawke in a scene from *Gattaca*. Image courtesy: IMDB.

Types of genetic testing

There are three types of genetic testing space agencies will use to select their ECM astronauts. The first of these – *diagnostic* testing – is used to identify or rule out a specific genetic or chromosomal condition. The second – *carrier* testing – will be used to identify candidates carrying one copy of a gene mutation that may cause a genetic disorder, while *predictive* testing will be used to detect gene mutations associated with disorders present in the candidate. This latter type of testing will identify those at risk of developing a disease, such as cancer, during a mission. Obviously, if any of the results are positive, the candidate will be eliminated from the recruitment process.

The tests described are performed on a sample of blood, hair, or skin, which is sent to a laboratory where technicians search for differences in chromosomes, DNA, or proteins. Due to the problems in interpreting genetic tests, space agencies will need to exercise particular care in determining the genetic profile of candidates. For example, a negative test result means the laboratory did *not* detect an abnormal gene, chromosome, or protein. However, although such a result may indicate a person is not affected by a particular disorder, it is possible the test missed a disease-causing genetic alteration. This is because some tests simply cannot detect *all* genetic changes associated with a specific disorder. To eliminate any ambiguity, space agencies will hopefully discard uninformative and inconclusive tests and conduct secondary tests. However, in the event of a positive result, the likely consequence for the candidate will be elimination from consideration as an astronaut. Needless to say, the effect of a positive result on candidates who have spent their professional lives accumulating the qualifications to become an astronaut will be upsetting to say the least. However, such testing will be necessary, given the potentially dire consequences of an astronaut being diagnosed with a critical illness during the mission.

Once a crewmember is pronounced radiation-resistant, found to have high bone density, and be genetically free of any future disease, they can move on to the final phase of medical selection: precautionary surgery.

PRECAUTIONARY SURGERY

Having run the gauntlet of medical poking and prodding, psychiatric assessment, and genetic testing, interplanetary astronauts will have their appendices removed. If you're thinking this is a bit extreme, consider the case of Russian surgeon Leonid Rogozov (Figure 3.4). In 1961, poor Rogozov [1, 2] was the only physician stationed on an isolated 12-man base in Antarctica when he developed appendicitis. What follows is an account of Rogozov's experience: an event that space agencies will want to avoid repeating at all costs!

On November 5th, 1960, the sixth Soviet Antarctic expedition sailed from Leningrad. After 36 days at sea, 12 expedition members landed on the ice shelf of the Princess Astrid Coast. Their task was to build an Antarctic polar base at Schirmacher Oasis and overwinter there. Nine weeks later, on February 18th,

50 Medical qualification for exploration class missions

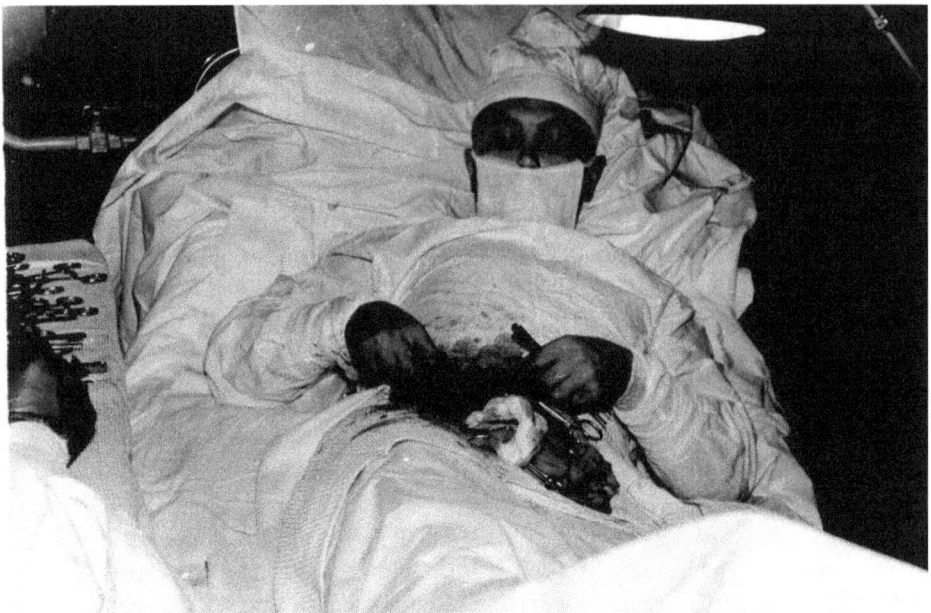

Figure 3.4 Leonid Rogozov performing his auto-appendectomy in the Antarctic. Rogozov first infiltrated the layers of his abdominal wall with 20 ml of 0.5% procaine, using several injections. After 15 min, he made a 10–12-cm incision. The visibility in the depth of the wound was not ideal, which meant he had to sometimes raise his head to obtain a better view, but for the most part, he worked by feel. After 30–40 min, Rogozov started to take short breaks because of general weakness and vertigo. Finally, he removed the severely affected appendix. He applied antibiotics and closed the wound. The operation lasted an hour and 45 min. Image courtesy: British Medical Journal.

1961, the new base, Novolazarevskaya, was declared operational. Shortly after, the Antarctic winter brought darkness, snowstorms, and froze the sea. Contact with the outside world was no longer possible. Like a crew of interplanetary-bound astronauts, the 12 expedition members of Novolazarevskaya had to rely on themselves.

One of the expedition's members was Leningrad surgeon Leonid Rogozov, who served as the expedition's doctor. Less than three months after arriving at Novolazarevskaya, Rogozov became ill. His symptoms included weakness, nausea, and pain in his stomach, prompting Rogozov to acknowledge he had acute appendicitis; if he was to survive, he had to undergo an operation – something that was impossible in the middle of the Antarctic night.

Rogozov treated himself using antibiotics, but his condition worsened. Outside, a snowstorm howled, shaking the base. After a sleepless night, Rogozov came to the only decision available: he would have to perform the operation. Following his instructions, the other crewmembers sterilized the bedding and instruments and prepared an improvised operating theater. Station director Vladislav Gerbovich selected the expedition's meteorologist Alexandr Artemev and the mechanic Zinovy

Teplinsky to assist in the operation. After undergoing a sterile wash, Rogozov explained the details of the operation and assigned his fellow crewmembers tasks: Artemev would hand him instruments, Teplinsky would hold the mirror and adjust the lighting, while Gerbovich was designated a backup in case one of the others fainted! In the event Rogozov lost consciousness, he instructed his team how to inject him with drugs and how to provide artificial ventilation. After briefing his assistants, Rogozov scrubbed and positioned himself before disinfecting and dressing the operating area. Anticipating the need to use his sense of touch to guide him, he decided to work without gloves.

The operation began at 2 am local time. Rogozov first dulled the affected area with procaine, a local anesthetic, using several injections. After waiting 15 min for the procaine to take effect, he made a 10–12-cm incision while Teplinsky held the mirror. The lighting and visibility were far from ideal and for most of the operation, Rogozov had to work by feel. During the procedure, Rogozov felt weak, forcing himself to take short breaks, which was why the operation took nearly two hours. After finally removing the appendix, he applied antibiotics and closed the wound. Later that night, Gerbovich wrote in his diary:

> "When Rogozov had made the incision and was manipulating his own innards as he removed the appendix, his intestine gurgled, which was highly unpleasant for us; it made one want to turn away, flee, not look – but I kept my head and stayed. Artemev and Teplinsky also held their places, although it later turned out they had both gone quite dizzy and were close to fainting ... Rogozov himself was calm and focused on his work, but sweat was running down his face and he frequently asked Teplinsky to wipe his forehead .... The operation ended at 4 am local time. By the end, Rogozov was very pale and obviously tired, but he finished everything off."

After closing the wound, Rogozov showed his fellow crewmembers how to wash and put away the instruments. Then, he took sleeping tablets and lay down for a rest. For the next few days, he continued taking antibiotics. After five days, his temperature was normal and after a week, he removed the stitches. Two weeks after the operation, Rogozov was able to return to his normal duties. He made the following remarks in his diary:

> "I worked without gloves. It was hard to see. The mirror helps, but it also hinders – after all, it's showing things backwards. I work mainly by touch. The bleeding is quite heavy, but I take my time – I try to work surely. Opening the peritoneum, I injured the blind gut and had to sew it up. Suddenly it flashed through my mind: there are more injuries here and I didn't notice them ... I grow weaker and weaker, my head starts to spin. Every 4–5 minutes I rest for 20–25 seconds. Finally, here it is, the cursed appendage! With horror I notice the dark stain at its base. That means just a day longer and it would have burst."

More than a year later, the Novolazarevskaya team left Antarctica and Rogozov returned to his work at the Department of General Surgery of the First Leningrad Medical Institute. He never returned to the Antarctic.

52 Medical qualification for exploration class missions

Rogozov's auto-appendectomy (the proper medical term for what he did) was probably the first such successful operation undertaken out of a hospital setting, with no outside help and without any other medical professional around. Rogozov's experience is a testament to the determination and human will for life, but it's not something to be repeated on an interplanetary mission!

Appendicitis

With an overall frequency in the general population of about 6–7%, appendicitis is one of our most common digestive tract surgical diseases. The appendix is a closed-end narrow tube attached to the first part of the colon. If the opening to the appendix becomes blocked or the fatty tissue in the appendix swells, bacteria, normally found within the appendix, may invade and infect the wall of the appendix. This infection results in *appendicitis*, to which the body responds by inflaming the appendix, which may ultimately lead to rupture, followed by spread of bacteria outside the appendix. Alternatively, the appendix may become perforated leading to an abscess or, in some cases, the entire lining of the stomach may be infected. Appendicitis may require the contents of the stomach to be drained through a tube passed through the nose. Needless to say, in the confined environment of a spacecraft in zero gravity, such a procedure would challenge even the most experienced surgeon! Perhaps the most feared complication of appendicitis is *sepsis*, a condition in which bacteria enter the blood and infect other parts of the body. Even on Earth, sepsis is considered a serious complication, but to an astronaut bound for Mars or returning to Earth, such a complication would be a death sentence.

These complications alone represent a powerful argument for removing the appendices of ECM crewmembers, but there are other factors to consider, such as diagnosing the condition – a procedure that would use vital medical consumables. For a crewmember suspected of suffering from appendicitis, the only possible diagnostic procedures available would be a urinalysis and an ultrasound procedure. In the event of complications, a computer tomography (CT) scan and abdominal X-ray would be unavailable due to the limited medical resources, although a laparoscopy could possibly be performed. However, a laparoscopy – a procedure in which a small fiber-optic tube with a camera is inserted into the abdomen through a puncture hole made in the wall of the stomach – requires a general anesthetic and would present a challenging procedure in zero gravity. Furthermore, even on Earth, appendicitis is often difficult to diagnose because other inflammatory problems can mimic the symptoms of the condition.

Appendectomy

Should a crewmember be correctly diagnosed with appendicitis, the next problem would be treatment, involving the removal of the appendix in a procedure known as an *appendectomy*. This requires the surgeon to make a 4–6-cm incision in the skin and layers of the abdominal wall, in the area of the appendix, and to remove the appendix. If an abscess is present, pus must be drained before the abdominal incision is closed. In recent years, laparoscopic surgery has been used to perform the procedure, but in zero gravity, the method would present risks.

PRE-MISSION MEDICAL SELECTION

ECMs will inevitably require a reassessment of pre-mission screening. The development of more sophisticated selection and de-selection criteria will be the first step in this reassessment. The next step will be deciding how to implement genetic screening and pre-emptive surgery. Some may argue a third step should be a renewed emphasis on psychiatric evaluation, but this is probably the least important selection criterion given the wealth of knowledge gleaned from the multi-year missions completed by Amundsen and co. Given these extraordinary medical selection criteria, those ultimately selected for an ECM will no doubt breathe a sigh of relief and look forward to mission training and the mission itself! However, for those with their sights set on Mars, the medical challenges will be far from over.

REFERENCES

[1] Rogozov, L.I. Self Operation. *Soviet Antarctic Expedition Information Bulletin*, **4**, 223–224 (1964).
[2] Rogozov, V.; Bermel, N. Auto-Appendectomy in the Antarctic: Case Report. *British Medical Journal*, **339**, b4965 (December 2009).

Section II

Exploration Class Medical Challenges

4

Radiation

"Space radiation has not been a serious problem for NASA human missions because they have been short in duration or have occurred in low Earth orbit, within the protective magnetic field of the Earth. However, if we plan to leave low Earth orbit to go on to Mars, we need to better investigate this issue and assess the risk to the astronauts in order to know whether we need to develop countermeasures such as medications or improved shielding. We currently know very little about the effects of space radiation, especially heavy element cosmic radiation, which is expected on future space missions and was the type of radiation used in this study."

Philip Scarpa, MD, NASA Flight Surgeon

It is more than 40 years since astronauts ventured beyond Earth's protective magnetic shield and traveled to the Moon. Although the Apollo missions subjected astronauts to space radiation, the short duration minimized the risk, but an exploration class mission (ECM) will subject astronauts to much longer exposure. In fact, astronauts will be in deep space for so long they will run the risk of suffering all sorts of illnesses, such as various cancers and degenerative tissue disorders. Of course, mission planners will do their best to provide countermeasures and a storm shelter, but even with the best protection, shielding crewmembers from the effects of deep space radiation may prove impossible. That's because interplanetary astronauts will be exposed to two types of nasty radiation capable of going right through the human body and tearing apart strands of deoxyribonucleic acid (DNA), the software of life that resides inside a cell nucleus. Once damaged, these cells simply lose the ability perform normally and to repair themselves.

RADIATION TYPES

There are two primary forms of hazardous space radiation particles. High-energy particles (protons) emitted by the Sun during intense flares (Panel 4.1) is one type. These flares, known as solar particle events (SPEs), move outward at millions of kilometers per hour and could strike an interplanetary spacecraft in days. The peak

> **Panel 4.1.** Solar flares
>
> The two largest SPEs observed were in August 1972 and October 1989. Such events are used to provide realistic estimates of the SPE environment that may be encountered during missions taking place during active solar conditions and may also assist in the prediction of SPEs. For example, the October 1989 SPE was predicted by the National Oceanic and Atmospheric Administration (NOAA) as a result of an X-ray burst occurring one hour prior to the SPE onset. Although the actual event was predicted successfully, the severity of the SPE was not forecast with much success. To provide crews with as much warning as possible, scientists will monitor the solar weather. One feature they will pay particular attention to will be sunspots, which form where intense magnetic fields twist and poke through the surface. Sometimes, when these field lines twist to the point of snapping, sunspots explode and release enormous amounts of stored energy and hot gas into the Sun's outer atmosphere. The resulting eruption is called a solar flare, which accelerates subatomic particles to near light speed and spews out ultraviolet (UV) and gamma-ray radiation into space. Occasionally, the flares are followed by a coronal mass ejection (CME), in which billions of tonnes of the Sun's plasma are ejected into space, traveling at more than 1,500 km/sec. Depending on how far from Earth the astronauts are, the radiation pulse from the flare may take anything from 8 to 20 or more minutes to arrive. With no atmosphere shielding the vehicle, the spacecraft will be bombarded by the radiation pulse, inflicting serious injury upon the crew.

incidences of these flare events tend to occur in 11-year cycles, but large flare events (Figure 4.1) and solar storms have also been unexpectedly observed during supposedly quiescent periods of the cycle; astronauts working on the surface of the Moon or bound for Mars would be as good as naked in the face of a solar flare event.

Cosmic rays, the other big radiation concern, originate from undetermined galactic sources and pose a greater long-term risk for cancer, cataracts, and other illness. That's because cosmic ray particles are more energetic than their solar cousins; the particles are basically atomic nuclei stripped of electrons, able to penetrate many centimeters of solid matter. Now, when astronauts are on the surface of a planet or a moon, they're protected against cosmic rays because planets and moons offer some natural protection. Even the Martian atmosphere, which is only about 1% as dense as Earth's, still manages to stop most of the solar particles, although it lets most of the cosmic rays through. But when astronauts are in deep space, they're attacked by both types of radiation coming at them from all directions. In fact, exposure is about twice as bad while traveling through space compared to being on the surface of Mars.

Measuring radiation 59

Figure 4.1 The Sun shows a C3-class solar flare (white area on upper left), a solar tsunami (wave-like structure, upper right), and multiple filaments of magnetism lifting off the stellar surface. Image courtesy: National Oceanic and Atmospheric Administration.

MEASURING RADIATION

So, just how damaging is this radiation? Well, before we discuss the actual damage that radiation can inflict upon the body, it's important to understand that radiation effects are cumulative and because the sources of radiation are variable, their relative risk is also unpredictable. Another variable, as we'll see later, is how organ tissues vary in their sensitivity to radiation exposure.

When measuring the amount of radiation absorbed by the body, scientists use a unit known as the gray (Gy[1]), one unit of which is equal to one joule of radiation energy absorbed per kilogram of tissue. To determine how energetic (and damaging) radiation is, the Gy is multiplied by a quality factor. The quality factor is a measure that takes into account the relative effectiveness of the radiation in producing the biological effect. For example, it is known ionizing radiations such as protons, beta particles, and energetic ions of heavier elements cause more biological damage than radiations such as X-rays and gamma rays. The more damaging radiations are said to have a relative biological effectiveness (RBE) greater than 1.0. The RBE is defined as the ratio of the dose of a particular radiation to that of the test radiation required to cause an equal biological effect. For example, if 1 Gy of beta particles kills the same number of blood cells as 2 Gy of X-rays, then the RBE of the beta particles is 2.0. If the 1.0 Gy of beta particles is multiplied by the RBE factor of 2.0, the biological equivalent dose is 2.0 Sieverts.[2] Sieverts are used because they are the Standard International (SI) unit of biological equivalent radiation dose. It sounds a little convoluted but it helps to understand the difference between the two, since the RBE concept is used in defining space radiation health.

RADIATION DAMAGE

Radiation sickness is usually associated with radiation doses greater than 1 Sv occurring within 24 hr. Depending on the exposure, symptoms can range from nausea and vomiting to hemorrhage and diarrhea (Table 4.1).

As Table 4.1 shows, the higher the radiation dose, the more severe the symptoms, the most common of which are nausea and vomiting. That's because organ systems are particularly vulnerable to the insidious effects of radiation exposure. Put simply, if too many cells of a certain tissue die, organ function will be compromised. For example, if cells lining the gastrointestinal tract die in sufficiently large numbers, the gut will be unable to absorb food or maintain electrolyte balance. This is the reason why, after suffering a large radiation dose, victims experience nausea and vomiting. However, cells do not have to die for organ function to be disrupted. Radiation may injure cells via many different pathways, depending on the sensitivity of a given tissue. For example, if full repair of cells fails, but not to the point of leading to the death of subsequent generations of cells, the damaged cells may survive and transform into cells that can become cancer precursors. Alternatively,

---

[1] The centigray (0.01 Gy) is known as the rad. To put this into some kind of perspective, a cancer patient receives approximately 60 Gy during a full course of therapy.

[2] Centisieverts (cSv, 0.01 Sv) are also known as rems. Rem, short for Roentgen Equivalent Man, is the radiation dose that causes the same injury to human tissue as 1 roentgen of X-rays. A typical diagnostic CAT scan, the kind you might get to check for tumors, delivers about 1 rem. To die, you'd need to absorb 300 rem, suddenly. The key word is *suddenly*. You can get 300 rem spread out over a number of days or weeks with little effect. Spreading the dose gives the body time to repair and replace its own damaged cells. But if that 300 rem comes all at once, as would occur during a solar storm, then the effect is deadly.

Table 4.1. Short-term effects on humans of severe radiation.[3]

Dose (rem.)	Probable physiological effects
10–50	No obvious effects, except minor blood changes
50–100	5–10% experience nausea and vomiting for 1 day. Fatigue, but no serious disability. Transient reduction in lymphocytes and neutrophils.[4] No deaths anticipated
100–200	25–50% experience nausea and vomiting for 1 day, followed by other symptoms of radiation sickness. 50% reduction in lymphocytes and neutrophils. No deaths anticipated
200–350	Most experience nausea and vomiting on the first day, followed by other symptoms of radiation sickness such as loss of appetite. Up to 75% reduction in all circulating blood elements. Mortality rates 5–50% of those exposed
350–550	Nearly all experience nausea and vomiting on the first day, followed by other symptoms of radiation sickness such as fever and emaciation. Mortality rates of 50–90% within six weeks. Survivors convalesce for about six months
550–750	All experience nausea and vomiting within four hours, followed by severe symptoms of radiation sickness. Death up to 100%
750–1,000	Severe nausea and vomiting may continue into the third day. Survival time reduced to less than three weeks
1,000–2,000	Nausea and vomiting within one to two hours. Always fatal within two weeks
4,500	Incapacitation within hours. Always fatal within one week

damaged cells may lose some functional characteristics, in turn leading to organ failure. Then, there's the risk to fertility.

Prolonged exposure to deep space radiation is likely to result in reduced fertility or transient or temporary sterility, lasting from several months to several years. In some cases, if the exposure is particularly prolonged or severe, the sterility may be permanent. For males, the radiation dose required to cause temporary sterility is between 0.5 and 4.0 Gy, although a single acute dose of 0.15 Gy has been reported to cause a decrease in sperm count. Temporary sterility may last from several months to several years. Doses in the range of 2.5–4.0 Gy may cause permanent sterility, but infertility may also be caused by low-dose-rate protracted exposure. Put it this way: by the time exploration class astronauts return from their missions, they will most likely be sterile.[5] This is why space agencies will offer astronauts of reproductive age the option of banking sperm or eggs.

---

[3] Table adapted from Nicogossian, A.; Huntoon, C. (eds). *Space Physiology and Medicine*, 3rd edn. Lea & Febiger, Philadelphia (1994).

[4] Lymphocytes are a type of white blood cell that produces antibodies to kill pathogens that invade the body. Neutrophils are one type of a white blood cell that assists the body's immune system to ward off disease.

[5] A radiation dose in the range of 6.0–20.0 Gy is required to sterilize females, whereas temporary sterility may occur at doses of between 1.25 and 3.0 Gy. Doses of 2.0–6.5 Gy are required to sterilize 5% of females for more than five years.

62 **Radiation**

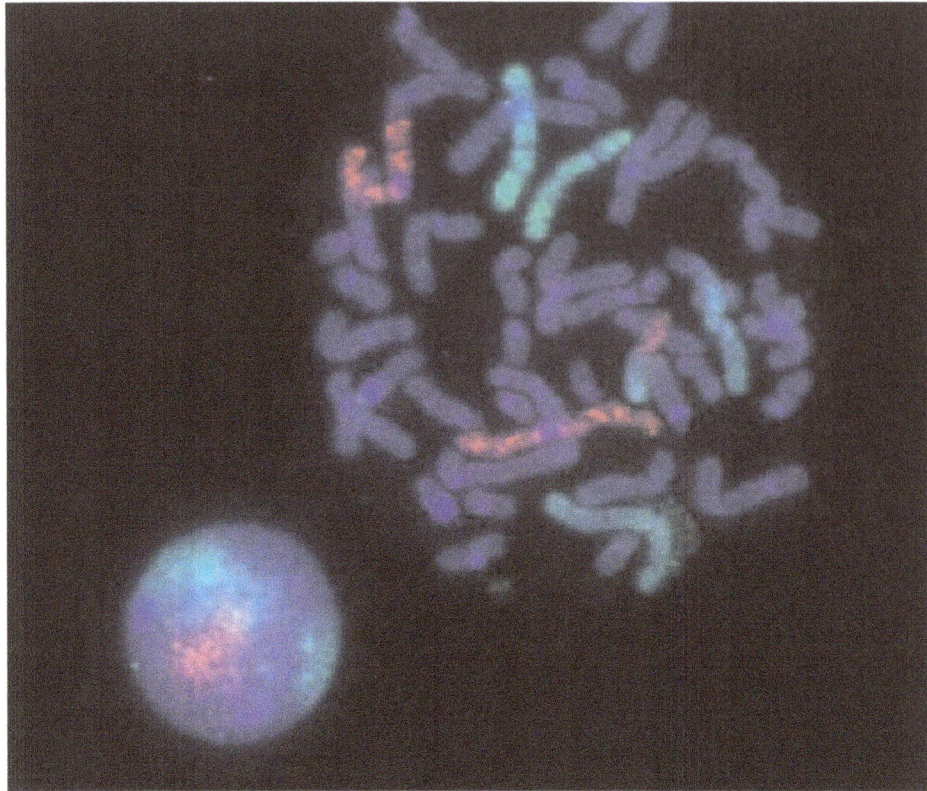

Figure 4.2 A blood sample from an ISS astronaut that has been damaged by space radiation. The strands are chromosomes "painted" with fluorescent dye. The picture shows big pieces of different colors stuck together. These are places where broken DNA has been repaired incorrectly by the cell. Image courtesy: NASA.

So far, we've discussed the immediate effects of a burst of radiation, but what about the late effects, which include induction of cancer, genetic mutations, and brain damage? We'll start by examining how radiation causes cancer.

When the human body is exposed to radiation, the energy from the radiation is deposited at the cellular level by interactions between the radiation and the electrons of molecules comprising the cells. The deposition of radiation causes the atoms that make up complex molecules to lose electron bonds that tie them to the molecule. In certain cases, the molecule will recover, but if the radiation deposition continues unabated, template molecules such as DNA may be unable to repair the damage and may die. Alternatively, cellular repair mechanisms may be unsuccessful and leave damaged DNA (Panel 4.2) cells incompletely repaired or lead to dying or aberrant cells in subsequent division. Such an unstable cell and its progeny will result in a little understood process known as *genomic instability*. Genomic instability is a hallmark of cancer cells and is thought to be involved in the process of carcinogenesis. In the hostile environment of deep space, radiation may induce such a process of

> **Panel 4.2. DNA mutation**
>
> DNA is the hereditary material in humans. Nearly every cell in a person's body has the same DNA. The information in DNA is stored as a code made up of four chemical bases: adenine (A), guanine (G), cytosine (C), and thymine (T). Human DNA consists of about three billion bases, and more than 99% of those bases are the same in all people. The sequence of these bases determines the information available for building and maintaining an organism, similar to the way in which letters of the alphabet appear in a certain order to form words. DNA bases pair up with each other, A with T and C with G, to form base pairs. Each base is attached to a sugar molecule and a phosphate molecule. Together, a base, sugar, and phosphate are called a nucleotide, arranged in two long strands that form a double helix. The structure of the double helix is similar to a ladder, with the base pairs forming the ladder's rungs and the sugar and phosphate molecules forming the sides.
>
> An important property of DNA is that it can replicate, or make copies of itself. Each strand of DNA in the double helix can serve as a pattern for duplicating the sequence of bases. This is critical when cells divide because each new cell needs to have an exact copy of the DNA present in the old cell. When these cells are damaged by radiation, mutations can occur. That's because radiation can damage DNA by altering nucleotide bases so that they look like other nucleotide bases. When the DNA strands are separated and copied, the altered base will pair with an incorrect base and cause the mutation. Radiation can also damage DNA by breaking the bonds – for example, breaking the phosphate backbone of DNA within a gene creates a mutated form of the gene and it is possible that the mutated gene will produce a protein that functions differently.

instability, resulting in the multiple gene mutations necessary for the development of cancer.

The damage to DNA (Figure 4.2) doesn't end there. In fact, in terms of heritable effects, the end points of exposure to radiation may include major congenital malformations, stillbirth, and tumors. Fortunately, some individuals possess genotypes that confer upon them an increased resistance to radiation. A genotype is simply a set of physical DNA molecules inherited from parents and to reduce the chances of genetic instability among its crewmembers, space agencies are sure to use genetic screening to select only those astronauts with reduced susceptibility genotypes. While the outlook for ECM astronauts may appear less than rosy, the repair mechanisms utilized by the human body to repair DNA after exposure to radiation are extremely versatile, and sophisticated cellular processes exist for repairing all types of DNA damage. These processes are capable of repairing not only base damage, but also single and double-strand breaks.

Just as troubling as the increased risk of cancer and the effect of radiation on DNA is the effect of heavy ions and the damage these particles inflict on the brain. In fact, heavy ions are emerging as one of the major hazards of interplanetary travel because they can inflict so much damage on the brain that astronauts could arrive at their destination only to find half their memory and learning capacity wiped out. That's because these particles can traverse several layers of cells and inflict not only cellular damage and biochemical changes, but also functional effects. In one computer-modeled estimate, 46% of the cells in the hippocampus (a center of memory and learning) would be struck by at least one heavy ion during a three-year trip to Mars. Because of the damaging effect of these ions, this would mean that 46% of the cells in the hippocampus would be destroyed. Research conducted by scientists at the NASA Space Radiation Laboratory (SRL) and at the Brookhaven National Laboratory (BNL) in Upton, New York, corroborated the computer model by studying mice. In the study, mice were administered a single dose of radiation equal to the amount astronauts might receive during a three-year return trip to Mars. Unexpectedly, the scientists found a special type of stem cell in the hippocampus is selectively killed by the radiation – a finding that prompted the following statement by Dennis Steindler:

> "The exceptional sensitivity of these neural stem cells suggests that we are going to have to rethink our understanding of stem cell susceptibility to radiation, including cosmic radiation encountered during space travel, as well as radiation doses that accompany different medical procedures."
>
> Dennis A. Steindler, Ph.D., Executive Director,
> McKnight Brain Institute, University of Florida, December, 2007

RADIATION EXPOSURE GUIDELINES

With so much radiation in space, you may be wondering what guidelines the space agencies have put in place to protect their astronauts. Well, according to National Council on Radiation Protection (NCRP) guidelines and the International Commission on Radiological Protection (ICRP), the maximum *annual* dose equivalent for the general public is one-thousandth of a Sievert (mSv). In contrast, the maximum annual dose equivalent for nuclear workers is 50 mSv. The maximum annual dose for astronauts in deep space? Well, to date, there are no guidelines for allowable radiation exposure in deep space. In a federally mandated obligation to follow the As Low As Reasonably Achievable (ALARA) principle, the guidelines NASA uses for its astronauts working on the International Space Station (ISS) are based on a point estimate[6] for the levels of radiation causing an excess risk of 3% for fatal cancer due to exposure (Table 4.2).

The best mission planners can do to predict radiation doses for a manned

---

[6] Point estimation uses sample data to calculate a single value, which serves as a "best guess" for an unknown population parameter.

Table 4.2. Radiation levels causing excess cancer risk.

(A) Recommended organ dose equivalent limits for all ages (NCRP)

Exposure interval	BFO[1] dose equivalent (cSv)	Ocular lens dose equivalent (cSv)	Skin dose equivalent (cSv)
30 days	25	100	150
Annual	50	200	300
Career	See Panel B	400	600

(B) LEO career whole-body effective dose limits (Sv) (NCRP)

Age	25	35	45	55
Male	0.7	1.0	1.5	2.9
Female	0.4	0.6	0.9	1.6

[1]Blood-forming organs.

interplanetary mission is to use the guidelines summarized in Table 4.2 and generate point estimates. One reason for this is because, computationally, the calculation of conventional exposures based on linear energy transfer (LET)[7] in a target medium such as the human body can be performed with little ambiguity. Unfortunately, this method, which is basically an advanced form of guessing, is fraught with uncertainty and any allowable doses calculated by this method can't be treated as a rigid requirement because the cell damage caused by ionizing radiation is highly variable for different cell types. Because of the uncertainty of extrapolating the LEO limits to deep space mission exposures, some scientists have devised a computation procedure, outlined here:

1. Divide the mission into phases and assign each phase a duration. For example, Phase One would be the transit phase from Earth to the destination.
2. For each phase of the mission, define the energetic particle fluences resulting from GCR and SPE.
3. Calculate the effect of the planetary atmosphere upon energetic particle fluences to define the radiation arriving at the surface.
4. Estimate the energetic particle fluence inside the planetary habitat.
5. Convert the net fluence into absorbed, equivalent, and effective doses and compare the estimated doses with permissible doses for astronauts.

Based on these computational procedures, it is estimated that during the 1,000-day mission duration of a manned Mars mission, the radiation doses will be in the order of 1,000 mSv, but these predictions don't consider the very real risk of solar flares.

---

[7] Linear energy transfer is a measure of the energy transferred to material as an ionizing particle travels through it. The measure is used to quantify the effects of ionizing radiation on the body.

COUNTERMEASURES

All this talk of genetic mutations and brain cells being killed suggests that perhaps astronauts will be risking too much by exposing themselves to deep space radiation but, while there is no doubt that radiation presents a looming hazard, it is far from being a showstopper. To protect astronauts, space agencies have devised a number of countermeasures that are usually classified into three categories: *operations*, *shielding*, and *biological*.

Operations

A unique aspect of an interplanetary mission is that there is no simple or fast abort-to-Earth option in the event of a radiation event, so mission planners can help reduce the risk of exposure by choosing more radiation-friendly trajectories. For example, the two principal classes of Mars mission trajectories are conjunction and opposition class. A conjunction class mission is considered a minimum delta velocity trajectory allowing 350–550 days on the surface of Mars and one-way trip times of between 200 and 300 days for a total round-trip time of between 900 and 1,000 days. Opposition class missions are characterized by short-duration stays of between 20 and 60 days, with one-way trip times of between 450 and 500 days. This latter class of trajectory is also characterized by asymmetric times between the inbound and outbound phases of the journey, with one phase being significantly longer than the other. Another aspect of this class of trajectory is a Venus swing-by, which would subject astronauts to more intense solar particle exposure. From the comparison of mission trajectories, it is clear that from a radiation exposure perspective, the long-stay fast transit trajectory is the safest option for a crewed Mars mission. Even safer is to use a faster means of propulsion such as the VASIMR mentioned in Chapter 2; such a system would ferry astronauts to Mars in just 37 days.

Shielding

A number of shielding options have been suggested ranging from cryogenic liquid tanks attached to the outside of the spacecraft to far-out concepts such as electric fields. One option is to use active shields that rely on magnetic or electric fields to deflect energetic particles, but designing a magnetic shield strong enough to deflect cosmic rays but weak enough to not harm astronauts is a challenge. For example, very large magnetic coils (longer than 10 km) have been proposed as a means of protecting astronauts, but such large coils produce instabilities that could damage the spacecraft. Also, such a system would require several billion volts to shield astronauts from the cosmic rays and so much energy just wouldn't be safe for the spacecraft. Other active shield concepts include plasma shields, but because these must produce several billion volts to protect against cosmic rays, they are unstable. In fact, such a concept results in stored energies equivalent to those from

Table 4.3. Chemical composition of candidate shielding materials.

Material	ID	Density (g/cm²)	Material	ID	Density (g/cm²)
Aluminum 2219	ALM	2.83	Lithium hydride	LIH	0.82
Polyetherimide	PEI	1.27	Liquid methane	LME	0.466
Polysulfone	PSF	1.24	Graphite nanofibers	GNF	2.25
Polyethylene	PET	0.92	Liquid hydrogen	LH2	0.07

nuclear weapon detonations, and require such a large structural mass to support the coils that the system exceeds the mass of traditional shielding! Also, if the coil is breached, an explosion of electromagnetic pulse would occur that would damage the spacecraft.

Since active shielding doesn't seem the way to go, what about the passive approach? Passive shielding involves quite literally placing shields around the spacecraft. It sounds simple, but because of the damaging effects of cosmic rays, passive shields must be very thick, which means a lot of weight. Another problem facing spacecraft designers is choosing a shield material that not only protects astronauts against radiation, but that also possesses the qualities required to build space structures. For example, materials with high hydrogen content generally possess high shielding properties but do not have the qualities required for building a spacecraft due to the lack of structural integrity of the material. Presently, the material candidates of choice amongst mission planners are organic polymers. Other multifunctional candidate materials being considered include liquid hydrogen, methane, and polyethylene (Table 4.3). These materials have been selected as candidate materials not only because of their shielding properties and structural integrity, but also because of their density – another important consideration when it comes to launching mission elements into orbit.

In addition to active shielding, the crew will also be able to make use of shielding provided by the structure of the spacecraft. This includes the avionics, life support, consumables, waste storage, and other hardware that protect crews from low-energy solar particles. A water bladder would also provide protection, but for the rarer, high-energy events, more protection is required. Since the duration of the most hazardous phase of an SPE or a close series of high-energy SPEs may last for hours or days, the spacecraft must be able to provide a storm-shelter capability for an extended period, during which the crew must have access to food, water, and hygiene facilities. Presently, the best shielding solution is to cover the spacecraft in 2.5-cm-thick slabs of high-density polyethylene – a strategy that will provide some protection, but not enough to defend against the effects of SPEs.

Biological

The third strategy being investigated to protect astronauts from the deleterious effects of radiation is the use of radioprotective agents. These agents are known as *radioprotectors* and protect cells from the damaging effects of exposure to ionizing radiation. Radioprotectors may be administered before and/or after radiation exposure and have been shown to work by a variety of mechanisms, such as their antioxidant properties. For example, research has shown amifostine and melatonin to be very effective as radioprotectors, although, at the time of writing, the drugs are only approved for clinical use. Amifostine, developed by the Walter Reed Institute (also known as WR2721), is the only radioprotectant approved for use in humans and has been shown to be effective when the drug is administered prior to an exposure. The drug has been on NASA's list of radioprotective drugs for several decades by virtue of its ability to inhibit the induction of mutations following radiation exposure, thereby reducing the risk of carcinogenesis. Another effective drug is genistein, which is an isoflavone (a type of antioxidant). Genistein is an ideal radioprotectant because it's non-toxic and is a natural product available in the diet from a single food source. It can also be taken daily by astronauts to provide a long window of protection. Studies have also shown genistein's efficacy in protecting against radiation-induced lethality and enhancing survival when administered one day before radiation exposure.

NANOTECH

Even by using advanced radiation shielding, radioprotectants, *and* optimum trajectories, it is unlikely astronauts will be completely protected from high-energy radiation. However, there is a near-horizon technology that offers a solution: nanotech [1]. Right now, scientists are designing microscopic vessels that can venture into the human body and repair problems one cell at a time (Figure 4.3). It may seem like a scene from the movie *Innerspace*: a tiny vessel, smaller than a human cell, tumbles through a patient's bloodstream, hunting down diseased cells and penetrating their membranes to deliver doses of medicine. But this isn't Hollywood. This is real science, funded by a grant from NASA. If successful, the nanobots (called nanoparticles or nanocapsules) developed by these scientists could protect astronauts from radiation damage.

Nanoparticles offer an elegant solution. Built using nanomaterials called dendrimers (Panel 4.3 and Figure 4.4), the molecule-sized sensors would be injected into astronauts to not only warn of radiation health impacts, but also to repair any damage caused by radiation. The drug-delivery capsules are tiny, measuring only a few hundred nanometers (a nanometer is one-millionth of a millimeter), which is smaller than a bacterium and smaller even than the wavelengths of visible light. A simple injection with a hypodermic needle could release millions of these capsules into the astronaut's bloodstream. Once there, nanoparticles would use the body's cellular signaling system to hunt down and repair radiation-damaged cells. How would it work? Well, the trillions of cells in the human body identify themselves and communicate with each other using molecules embedded in their membranes. These molecules act as chemical "flags"

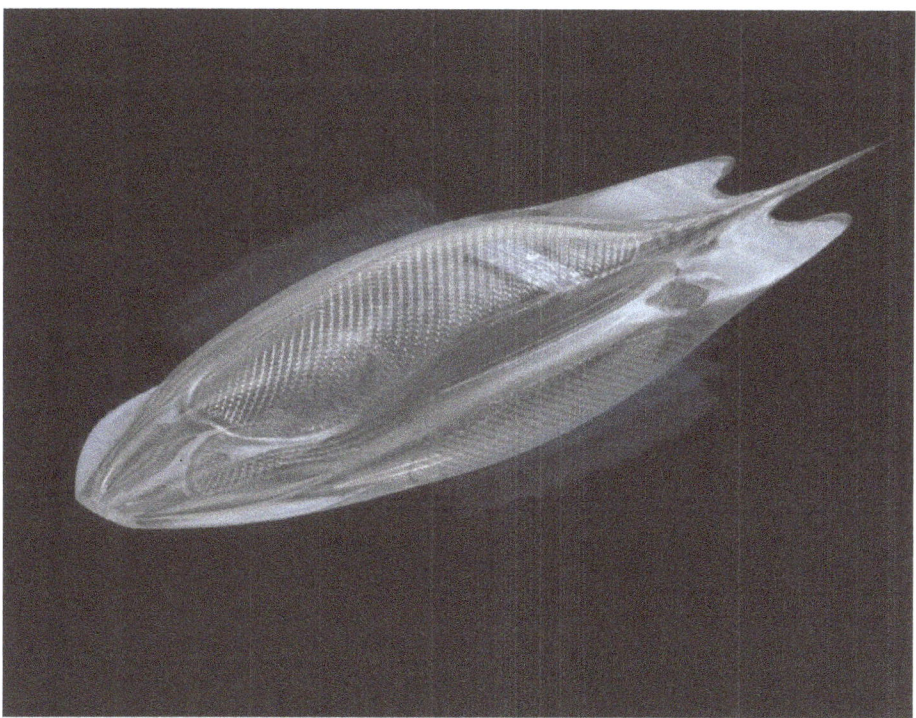

Figure 4.3 Cell rover. These nanobots patrol the circulatory system, searching for breaches. © E-spaces and Robert A. Freitas Jr, 3danimation.e-spaces.com and www.rfreitas.com and Philippe Van Nedervelde.

Panel 4.3. Dendrimers

The word "dendrimer" comes from the Greek "dendros", meaning "trees". Think of a tree in which each of its branches divides into two new branches after a certain length. This continues repeatedly until the branches become so densely packed that a canopy forms a globe. In a dendrimer, the branches are interlinked polymerized chains of molecules, each of which generates new chains, all of which converge to a single focal point or core.

Because dendrimers are such precisely defined chemical structures, they are the ideal building block for creating a biologically active nanomaterial.

---

for communicating with other cells and when cells are damaged by radiation, they produce markers and place them on their outer surfaces. Basically, it's a system whereby cells can talk to each other and say "Hey, I've been damaged". What nanoscientists would do is implant molecules on the outer surface of the nanoparticles that bind to the markers and then program the nanoparticles to

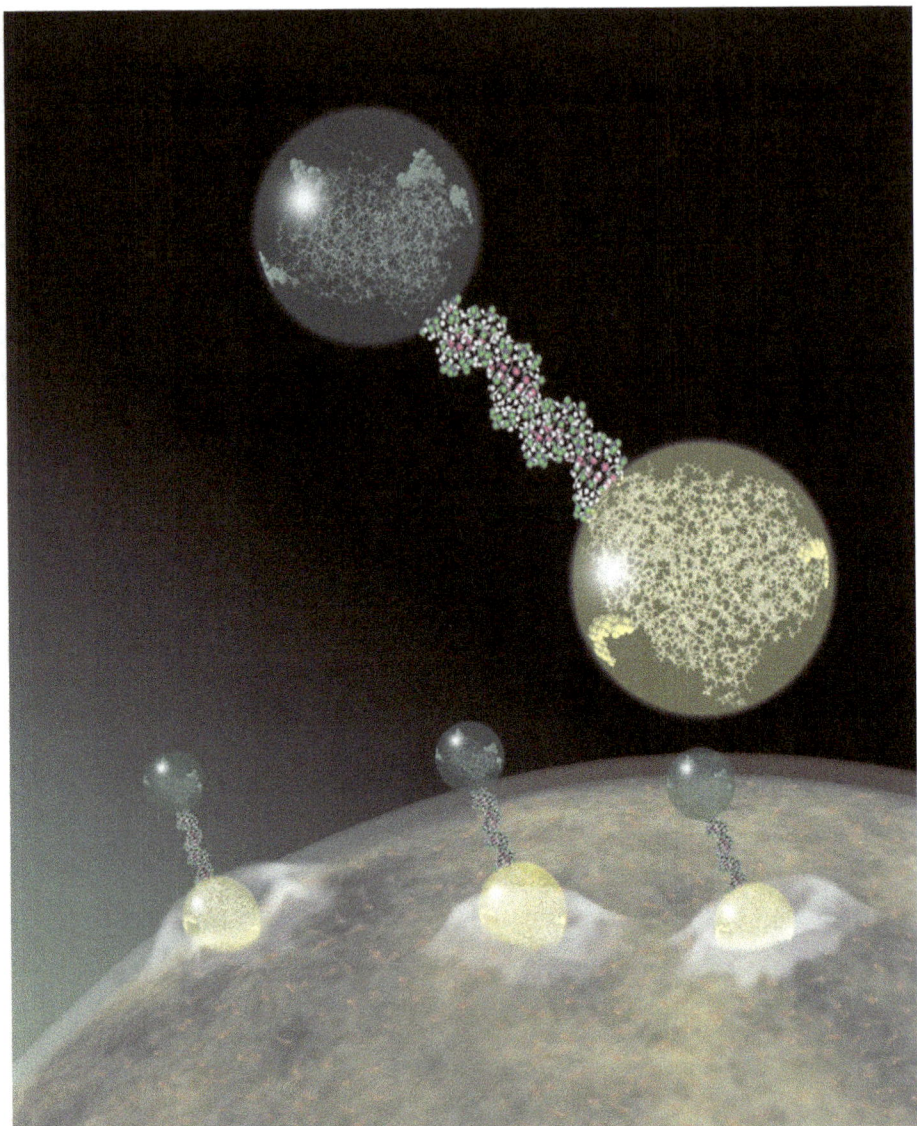

Figure 4.4 Dendrimer complex docking on cellular folate receptors. © E-spaces and Robert A. Freitas Jr, 3danimation.e-spaces.com and www.rfreitas.com and Philippe Van Nedervelde.

search for radiation-damaged cells. Once the nanoparticles find the damaged cells, they would assess the extent of the injury. If the cell was very badly damaged, the nanoparticles could enter the cell and program it for destruction. If the damage was judged repairable, the nanoparticles would release DNA-repair enzymes and fix the cell. The nanoparticles would also have fluorescent tags attached, which would provide a means of monitoring the work of nanoparticles inside the body.

While the nanosensors [2, 3] in the bloodstream would help astronauts to monitor radiation damage and even repair it, there's still a chance that even a system as versatile as this could be overwhelmed by a solar flare event. Such an event could put vital blood-making cells in jeopardy and without a fresh supply of red and white blood cells, astronauts would quickly become anemic, their immune system would collapse, and without medical attention, they would die. So why not simply replace the blood with a more rugged system? Well, that's exactly what nanoscientists Robert Freitas and Christopher Phoenix propose doing. Freitas and Phoenix's concept involves exchanging a person's blood with 500 trillion oxygen and nutrient-carrying nanobots [2, 3]. The system is called the vasculoid (a vascular-like machine), and it is designed to duplicate every function of blood, albeit more efficiently.

One of the key elements of the vasculoid is the respirocyte (Figure 4.5), a type of nanobot that is a key component of this nanotech blood. Each respirocyte is constructed of 18 billion precisely arranged atoms and has an onboard computer, powerplant, and molecular pumps capable of transporting oxygen and CO_2

Figure 4.5 Respirocyte in a blood vessel surrounded by red blood cells. The respirocyte is a nanobot capable of duplicating all thermal and biochemical transport functions of blood. © E-spaces and Robert A. Freitas Jr, 3danimation.e-spaces.com and www.rfreitas.com and Philippe Van Nedervelde.

molecules. Not only is it capable of duplicating all thermal and biochemical transport functions of blood, it is also able to perform these functions several hundreds of times more efficiently than biological blood. In essence, the vasculoid is nothing short of a mechanically engineered redesign of the human circulatory system. Despite the complexity of the system (it comprises 500 trillion independently cooperating nanobots), it weighs only 2 kg and is powered by nothing more than glucose and oxygen.

The key structural element of the vasculoid is a two-dimensional vascular-surface-conforming array of 150 trillion square plates, called sapphiroids. The sapphiroids are self-contained super-thin nanorobots that cover the entire surface of all blood vessels in the body, to one-plate thickness. Of the 150 trillion plates, 24 trillion are molecule-conveying docking bays where tankers containing molecules for distribution can dock and load/unload their cargo. Another feature of the array is the cellulock, of which there are 32.6 billion. At the cellulocks, boxcars carrying biological cells dock and load or unload their cargo. The remaining 125 trillion plates are reserved for special equipment and other applications. All the plates have watertight mechanical interfaces comprising metamorphic bumpers along the perimeter of each plate, which allow the bumper to expand and contract in area. It's a feature that permits the system to flex in response to body movements.

A discussion of all the vasculoid's components is beyond the scope of this chapter, so we'll focus on how the vasculoid would help astronauts suffering from radiation sickness. Remember, ionizing radiation causes atoms and molecules to become ionized or excited. These excitations and ionizations can produce free radicals, break chemical bonds, and damage molecules that regulate vital cell processes. Although regular cells can repair certain levels of cell damage, at higher levels, cell death results and at extremely high doses, cells can't be replaced quickly enough, and tissues fail to function. If this were to happen in a vasculoid-installed astronaut, the vasculoid would detect the damage and deploy armies of vasculocytes (Panel 4.4). Vasculocytes are independent nanobots equipped with ambulatory appendages, manipulator

---

Panel 4.4. The vasculocyte

The vasculocyte is a squat, hexagonal-shaped nanorobot with rounded corners, measuring 2.7 microns across and 1 micron tall, which means it's so small that it easily fits inside the narrowest capillaries of the human body. Its 400-billion-atom structure weighs about 8 picograms. On its six side walls, the vasculocyte is covered by an extensible "bumper" surface that cycles between 100 and 300 nm of thickness as internally stored ballast water inflates and deflates the surface every second. This cycling allows the vasculocyte sitting on a blood vessel wall to continuously adjust its girth to match the changes in blood vessel circumference that occur during each systolic pulse of the heart.

arms, repair and assembly tools, onboard computers, communications, and independent power supplies. They patrol the vasculoid continuously, searching for maintenance and repair tasks such as plugging internal leaks, cleaning spills, and leak scavenging. In the event of radiation damage, these nanorobots would search out affected cells using certain molecular markers. They would then destroy these cells, and only these cells.

Now, installing a system as complex as the vasculoid might sound like a risky proposition. Surely, something so complicated can't be reliable? Not so, say the designers, who seem to have thought of everything; to allay fears of system failure, the vasculoid's major subsystems incorporate 10-fold redundancy (by comparison, many of the systems in the Space Shuttle have three levels of redundancy).

Installation

By now, you may be wondering how an astronaut could be fitted with such a device. Unlike donor organs, which are implanted, the vasculoid would be installed[8] in a complex process that begins with exsanguination and finishes with the intricate vascular plating operation mentioned earlier. After being sedated, the astronaut's natural circulatory fluids would be removed and replaced with installation fluids. This step would be followed by mechanical vascular plating, defluidization, and finally activation of the vasculoid and rewarming the astronaut. From start to finish, the installation process would take about six hours. What follows is the step-by-step process as it may occur in the near future.

An astronaut being installed would be informed they are about to undergo a major medical procedure that involves replacing about 8% body mass with complex nanomachinery. Such an operation is not without risk, so the astronaut would receive psychological counseling to deal with the personal implications. Preparation would begin 24 hr before installation, when the astronaut would receive an injection of 70 billion vascular repair nanorobots. These mobile, artery-walking nanobots would clean out any fatty streaks, plaque deposits, lesions, infections, and vascular wall tumors.

After completing their tasks, the repair devices would be exfused and the results downloaded to a computer. This information would be used by the surgeon to prepare a map of the patient's vascular tree to improve efficiency during plating and plate initialization. Once this step is complete, the astronaut would be sedated, cannulated, and hooked up to a heart–lung machine. Heparin and streptokinase

---

[8] This might sound far-fetched, but scientists at Albert Einstein College of Medicine of Yeshiva University have already infused nanoparticles into mice. Research has shown that melanin protects against radiation by helping to prevent the formation of free radicals, which cause DNA damage. To provide protective melanin to the bone marrow, scientists created melanin nanoparticles and injected them into the bone marrow of mice.

would be injected to prevent clotting, after which the surgeon would administer various agents to aid the installation process.

After the astronaut had been anesthetized, their entire blood volume would be replaced with a suspension of respirocytes and a mixture of electrolytes and other components normally found in blood substitutes. The respirocyte fleet would provide oxygen and carbon dioxide transport equivalent to the entire human red blood cell (RBC) mass for three hours after the cessation of respiration. Once the blood volume had been exchanged, the astronaut's core temperature would be reduced from 37°C to just 7–17°C, after which the astronaut would be ready for the intravenous deployment of vasculoid components. First, the respirocyte suspension would be replaced by a new suspension containing 1% fully charged respirocytes and 10% cargo-bearing vasculocytes, creating a mixture whose viscosity and flow characteristics approximate to human blood. Each vasculocyte would drift in the flow until it encountered a vessel wall, which would activate it, causing it to release its cargo. If the immediate area was already plated, the vasculocyte would simply walk across the surface until it reached a clear area to deposit its cargo. Once its cargo plate was in place, the vasculocyte would release back into the fluid, power down, and be exfused from the body. After positioning and subsystem validation, each plate would inflate fluid-tight metamorphic bumpers along its contact perimeter with its neighbors, which would lock their bumpers firmly together with reversible fasteners embedded in the bumpers. After about an hour, the structure of the vasculoid would be almost complete and all the major components would have been tested. The astronaut would now be ready for defluidization.

During the defluidization stage, a monolayer of nanorobotic plates would form a chemically inert, flexible sapphire liner on the vascular tree's interior surface and vasculo-infusant fluid would be purged from the body by introducing 6 liters of oxygenated acetone to rinse the vascular tree. Once the system had been rinsed, the process of plate initialization would begin. With 200 billion vasculocytes and 150 trillion plates to initialize, each active vasculocyte would need to contact and initialize 750 plates. This stage would be followed by the installation of storage vesicles that contain reserves of mobile and cargo-carrying nanodevices and other auxiliary nanodevices. The astronaut would then be rewarmed, catheters would be removed, and the vascular breaches sealed. At this stage, the vasculoid would be operational and essential metabolic and immunological systems would have returned to normal. While the vasculoid is intended to be a permanent feature, it is probable that crewmembers would want the device extracted on their return from their mission.

The installation of such a device into an astronaut for the purpose of protecting crewmembers from radiation damage may represent one of the most extreme medical interventions that will only be possible if significant advances in medical molecular nanotechnology are realized. However, current knowledge of nanomechanical systems suggests that such a device would not violate known physical, engineering, or medical principles and could be made safe for the user. If, in fact, the vasculoid becomes a reality, it may represent a significant outpost not only in biological evolution, but in Man's quest to extend the frontier beyond orbit.

REFERENCES

[1] Freitas Jr, R.A. Respirocytes: High Performance Artificial Nanotechnology Red Blood Cells. *NanoTechnology Magazine*, **2** (1), 8–13 (October 1996).
[2] Freitas Jr, R.A. Exploratory Design in Medical Nanotechnology: A Mechanical Artificial Red Cell. *Artificial Cells, Blood Substitutes, and Immobil. Biotech.*, **26**, 411–430 (1998).
[3] Freitas Jr, R.A. Robots in the Bloodstream: The Promise of Nanomedicine. *Pathways, The Novartis Journal*, **2**, 36–41 (October–December 2001).

5

Bone loss

While radiation damage may be successfully overcome thanks to nanotech, the problem of bone loss is one that still concerns scientists. There's no doubt that astronauts experience greatly increased rates of bone loss when exposed to reduced gravity environments, but why this happens still has researchers somewhat baffled. Is the mechanical signal of gravity somehow converted to a chemical signal that regulates bone growth or is the loss caused by a decrease in formation rates or an increase in degradation rates? And to what extent do the rapid rates of degradation continue? Answers to these and other questions remain, which is why bone loss remains one of the most significant obstacles to long-duration spaceflight:

> "Bone demineralization during the Mars stay is unknown, but should be less than for an equivalent 0 g exposure. During the transfers the level of demineralization could reach 50% at the pelvis and it will certainly be more than the 15% threshold (considered as the level of significant increase of bone fracture risk). Bone demineralization is therefore an unacceptable risk, and must be controlled."
>
> Statement from the paper Definition of Reference Scenarios for a European Participation in Human Exploration and Estimation of the Life Sciences and Life Support Requirements, European Space Agency, September 2000

Bone demineralization begins immediately on arrival in space. During the first few days of a mission, a 60–70% increase in the amount of calcium excreted by the body is observed. The loss is rapid and continuous, resulting in losses of bone mineral, changes in bone architecture, and alterations in skeletal mass that result in a condition similar to osteoporosis. This microgravity-induced loss of bone mineral density (BMD) has been documented primarily in the weight-bearing components of the skeletal system, such as the lumbar vertebrae, femoral neck, trochanter, tibia, and calcaneus. Research onboard the International Space Station (ISS) indicates astronauts may lose between 1 and 2% of their BMD per month – a rate almost five times the rate of women with postmenopausal osteoporosis! So, imagine a crew en route to Mars. After spending six months traveling to the Red Planet and six months exploring its surface, astronauts may lose 20% of their BMD, equating to a 40% loss in bone strength. In fact, the loss of trabecular bone could be so great that the body would be unable to rebuild the bone architecture on return to Earth!

Although the reduced gravity of Mars and other interplanetary destinations will lessen the effect of bone demineralization, the sheer magnitude of bone loss will mean astronauts will still be highly susceptible to the risk of fracture. Furthermore, in the event of a crewmember suffering a fracture, healing would be inhibited due to the reduced gravitational field of whichever planet they were exploring. Then there are the problems of astronauts returning from interplanetary missions never fully recovering their bone mass and the related health hazards such as toxic accumulations of excess mineral in the kidneys. It's a major concern among mission planners, which is why scientists are working diligently to find a solution, but before we discuss possible countermeasures, it's useful to understand the hazards presented by bone demineralization and the physiological processes that occur in the skeletal system during exploration class missions (ECMs).

EFFECT OF MICROGRAVITY ON THE SKELETAL SYSTEM

One of the most regularly documented physiological changes associated with the spaceflight environment is the process of bone demineralization, caused by the absence of weight bearing. An absence of load removes not only the direct compressive forces on the long bones and spine, but also the indirect loading on these bones from the pull of muscles on the various bone structures to which they are attached. Invariably, the unloading of the skeleton leads to osteoporosis (Figure 5.1), weakening of the bones, and delayed healing of fractures.

Bone is composed of mineral and organic components. Collagen, the most abundant protein in bone, is synthesized primarily by osteoblasts (bone cells responsible for removing bone tissue) and forms a framework upon which mineralization is superimposed. Adding to this process are various matrix proteins that have cell recruitment functions in remodeling bone. At present, there is little information concerning the influence of zero gravity on the biophysical functions of these matrix proteins. Compounding the issue is the suggestion of an impaired mineralization process that may occur during spaceflight. You see, bone demineralization is a complex and dynamic sequence of events involving mineral deposition regulated by cells responsible for aligning calcium phosphate crystals and depositing them within the collagen structure. Evidence from spaceflight indicates these minerals, when formed in microgravity, have a decreased crystal size and are configured imperfectly.

Also important to the understanding and prevention of bone loss during zero gravity are the processes of repair and remodeling. The remodeling process is governed by osteoblasts and osteoclasts (bone cells responsible for bone repair), although the control mechanism has not been identified. It is known that in space, the astronaut's skeleton undergoes a fast rate of resorption due to the unloading of mechanical stresses and weights. As a result of the skeleton no longer having to bear

Effect of microgravity on the skeletal system 79

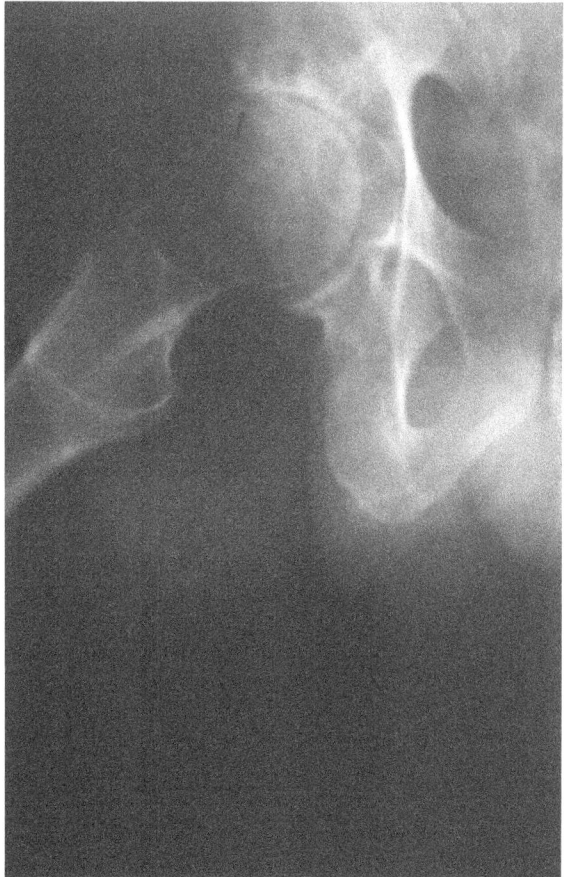

Figure 5.1 Osteoporosis is an occupational hazard of long-duration astronauts. The gray area in this image of the pelvis and femur shows where bone has wasted away. Image courtesy: NASA.

the astronaut's full weight, the body signals to the osteoclasts to resorb bone at a fast rate and thereby begins to rid itself of what it believes is unnecessary bone. This process occurs in tandem with the rate of bone formation, which is negatively affected by zero gravity, resulting in a slowing of the action of osteoclasts and reducing the amount of calcium absorption. For example, on Earth, bone absorbs 40–50% of the calcium intake, whereas only 20–25% is absorbed in space.

Another component of the bone loss mystery is the process of bone homeostasis. Bone tissue is constantly recycled and renewed to maintain homeostasis – a process of bone remodeling and repair resulting in approximately 500 mg of calcium entering or leaving the bone each day. This remodeling occurs selectively in a process of reabsorbing or depositing bone tissue determined by the mechanical or gravitational stresses acting on the bone. Together, the osteoclast and osteoblast cells remodel bone tissue continuously in a process controlled by hormonal and mechanical feedback. Unfortunately, removal of gravitational stress results in a disruption to both of these feedback processes, resulting in bone wasting away – a situation compounded by the effect of the blood supply.

We know that all the physiological processes in bone depend on an optimal bone blood supply, but to understand how blood supply can cause bone atrophy, we must also be familiar with the effect of the absence of gravity on the circulatory system. Gravity affects an organism *hydrostatically*, so when an astronaut on Earth is in an upright position, the proportion of fluid volume in their lower half is greater than in their upper half. But, once the force of gravity is removed, the hydrostatic forces exerted on bodily fluid are completely neutralized and blood is distributed evenly throughout the body. This means the body detects less blood in the extremities such as the legs and the body's response to this unnatural blood redistribution is to pump more blood through the heart. Unfortunately, the increase in blood circulation leads to *accelerated* demineralization because increased blood flow results in an increased blood velocity through bone, which increases the rate of calcium absorption into the blood supply. It really is a lose–lose situation, but there's more bad news!

As if the altered physiological processes weren't bad enough, bone demineralization is exacerbated by the effect of radiation, which results in a condition known as osteoradionecrosis. This is a condition of non-living bone in a site of radiation injury. It's been observed in cancer patients receiving high doses of radiation during chemotherapy. Although research has not investigated the effect of ionizing radiation on general bone quality, there is a high risk that interplanetary astronauts may be exposed to sufficient radiation to cause significant decreases in both bone volume and bone integrity. To assess the effect of radiation upon bone architecture during long-duration missions, a recent study used microcomputed tomography to measure the effects of whole-body exposure to space-relevant radiation in mice. Conducted at Clemson University, South Carolina, and the Brookhaven National Laboratory (BNL), the study subjected groups of mice to radiation similar in intensity to that which interplanetary astronauts might experience. Four months after exposure, the left tibiae and femurs were removed and analyzed by microcomputed tomography to measure parameters such as bone volume and connectivity density. The results of the Clemson study were quite alarming because some of the changes suggested permanent deficits in bone integrity and the reduced ability of the bone to sustain loading. It was suggested that although bone that had been exposed to space-relevant radiation might recover bone mass, the ability and the efficiency of the bone to transmit loads may be permanently compromised.

With all these problems associated with bone loss, it will be important for ECM crews to monitor their rate of bone loss and BMD. Some of the current methods of measuring BMD include ultrasound, computed tomography, magnetic resonance imaging (MRI), and Dual-Energy Absorptiometry (DEXA). Of these, perhaps the most accurate is DEXA, a system in which two low-dose X-ray beams of different energies are used to scan regions of the body suspected of bone loss. The reason two different X-ray energies are used is to distinguish between bone and muscle, since each tissue absorbs differently. Although the results from a DEXA scan provide a reasonably accurate determination of BMD, one of the drawbacks of the system is

Effect of microgravity on the skeletal system 81

Figure 5.2 The AMPDXA equipment in a clinical setting. A scaled-down version of this may be used by astronauts en route to Mars to assess bone integrity. Source: Charles, H.K., Jr; Chen, M.H.; Spisz, T.S.; Beck, T.J.; Feldmesser, H.S.; Magee, T.C.; Huang, B.P. AMPDXA for Precision Bone Loss Measurements on Earth and in Space. *John Hopkins APL Technical Digest*, **25**(3), 192 (2004).

its inability to distinguish between compact and cancellous, making it is almost impossible to reconstruct an engineering model of the bone to perform the necessary stress-loading simulations. Since it is necessary to determine the specific location of bone loss to accurately assess fracture risk, a more sensitive means of assessing BMD is required. For interplanetary missions, this equipment also needs to be flight-qualifiable. Such a system might be a scaled-down version of the Advanced Multiple-Projection Dual-energy X-ray Absorptiometry (AMPDXA) system, which allows a much higher-resolution image to be produced (Figure 5.2).

Because the system uses multiple images acquired at different angles, it is possible to determine precise BMD and bone geometry images that may be used for fracture assessment and thereby permit longitudinal studies of bone in space. By using this system, astronauts and ground-based flight surgeons will be able to accurately monitor rates of bone loss, but while having the ability to monitor bone loss will undoubtedly be helpful, the availability of a measuring system does nothing to reduce the loss of bone. To achieve this, countermeasures will need to be implemented.

COUNTERMEASURES TO BONE DEMINERALIZATION

The most common countermeasures to bone demineralization may be broadly classified into pharmacological intervention and non-pharmacological intervention. Pharmacological intervention includes the use of osteoporosis drugs such as alendronate (marketed under the brand name Fosamax™) and *calcitonin* (marketed under the brand name Miacalcin™). Both of these are approved by the Federal Drug Administration (FDA) but unfortunately for astronauts and flight surgeons, while the claims made for their efficacy suggest they have potential in bone-loss prevention, there are a number of drawbacks to their use. For example, alendronate, while very effective in promoting bone mass, must be taken for several years to gain the maximum benefit and the side effects of such long-term use are completely unknown. The other option, calcitonin, is a hormonal drug resulting in bone mass gains of only 1½% a year – a figure far short of the required gains needed to offset losses during a multi-year mission. A more controversial drug is Slow Release Sodium Fluoride (SRSF), a formula that boosts the efficiency of bone-building osteoblasts but requires patients to have an annual blood fluoride check to ensure the drug stays below toxic levels in the body!

While calcitonin, alendronate, and SRSF may not be the drugs of choice for interplanetary astronauts, a more promising formulation is Osteoporex™, a unique sea-algae calcium that is 90% absorbable by the body. Backed by more than a decade of research involving more than 300 treatment studies, the supplement has proved successful in promoting bone mass in 95% of the studies. An all-natural nutrient supplement that is four times as effective as synthetic pharmaceutical drugs, Osteoporex™ may, in conjunction with other countermeasures described here, prove to play an important role in keeping ECM astronauts' bones strong.

Now, you might be wondering why astronauts simply don't increase their calcium intake. After all, part of the reason bone mass is lost is because calcium is lost from the body. Well, it may seem a simple solution to the problem, but the remedy is a little more complex. You see, simply adding more calcium to the astronaut's diet wouldn't help and might even make the problem worse because excessive dietary calcium disrupts the delicate mineral balance needed by the body to repair and build bone. When the body's mineral content is over-weighted in favor of one particular mineral, the vital mineral balance is thrown off and it becomes more difficult to utilize *any* of the minerals properly. Because of this, researchers have directed their attention on proper calcium absorption and have discovered calcium balance can be maintained if calcium is used in small doses, in a highly absorbable form, and in proper balance with other absorption-promoting nutrients that enhance calcium metabolism. However, while this might help, it still won't be enough to help astronauts embarked upon multi-year missions. To further protect crewmembers, it will also be necessary to use non-pharmacological intervention strategies.

On Earth, non-pharmacological strategies such as exercise combined with adequate calcium, Vitamin D, and protein intake will maintain and even increase bone mass; a daily calcium intake of 0.1–1.0 g/day combined with a Vitamin D intake of 800 IU/day has proven to reduce the risk of fracture. But, for interplanetary astronauts, more aggressive supplementation will be required. For example, on Earth, humans need about 200 IU/day of Vitamin D for bone growth, which most people achieve from exposure to sunlight. However, astronauts will be stuck inside a spacecraft and will need to maintain their Vitamin D levels by other means. Studies conducted onboard nuclear submarines suggest that in the absence of exposure to sunlight, Vitamin D intake should be boosted to between 500 and 600 IU/day. Another method by which Vitamin D levels may be maintained is by providing astronauts with a lighting system emitting the required amount of ultraviolet (UV) radiation to stimulate Vitamin D production.

In addition to taking drugs and increasing their calcium and Vitamin D intake, astronauts may benefit from eating fish oil. In 2010, NASA scientists found that by adding an omega-3 fatty acid called eicosapentaenoic acid (EPA) to regular bone cell

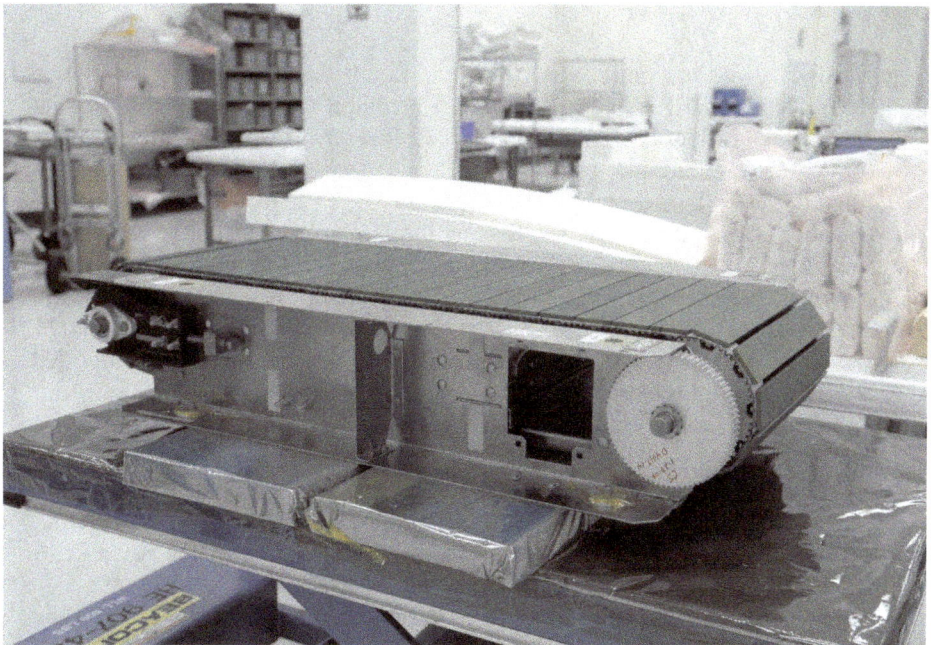

Figure 5.3 The Combined Operational Load-Bearing External Resistance Treadmill (COLBERT) is adapted from a regular treadmill. Engineers designed the device to allow astronauts to run on it without shaking the International Space Station or disturbing experiments taking place. Image courtesy: NASA.

cultures, the activation of factors that lead to bone breakdown was inhibited. The key factor that leads to bone loss (and muscle loss) is known as "nuclear factor kappa B" (a protein complex that controls the binding of DNA) or NFKB. Based on their findings, the scientists evaluated bone loss in astronauts and compared the results to reported fish intake during spaceflight. It was found that astronauts who ate more fish lost less bone mineral after four-to-six-month spaceflights. While it may be premature to conclude that the solution to the problem is simply a matter of diet, the scientists did find there was a link between the numbers of times astronauts ate fish in flight and the amount of bone they lost after flight.

While drugs and nutrition will no doubt help astronauts maintain their bone strength, there is another countermeasure that has been used almost as long as there have been astronauts: exercise. You've probably seen astronauts running on treadmills onboard the ISS and if you're a fan of Stephen Colbert, you'll probably remember the fanfare accompanying the installation of the new treadmill (Figure 5.3) named after the talk-show host (Panel 5.1).

It remains to be seen whether a version of the COLBERT will be included in the exercise toolkit of a future interplanetary spacecraft, but there will almost certainly be a treadmill together with a myriad other exercise devices. In common with current

---

Panel 5.1. The COLBERT

Officially called the Combined Operational Load-Bearing External Resistance Treadmill (COLBERT), the ISS's new $5 million treadmill actually got its name as a consolation prize for Stephen Colbert, who won an online NASA contest for the naming rights to a new space station module. In the competition, Colbert destroyed the competition, with the "Colbert" suggestion accumulating 230,539 votes, beating the runner-up, "Serenity", by 40,000 votes. But NASA ignored the votes (so much for fair competition!) and decided to name the module "Tranquility". As Bill Gerstenmaier, associate administrator for Space Operations, explained in a press release: "Apollo 11 landed on the Moon at the Sea of Tranquility 40 years ago this July. We selected 'Tranquility' because it ties it to exploration and the moon, and symbolizes the spirit of international cooperation embodied by the space station."

Despite the boos and hisses that poured out from the audience when the decision was announced, Colbert was soon appeased when astronaut, Sunita Williams, told him his name would grace a very important piece of equipment: the COLBERT treadmill!

The commercial off-the-shelf space-modified treadmill was unpacked on the orbiting station in September 2009 and took astronauts more than 20 hr to put together from more than 100 pieces. After conducting a series of tests to make sure it was working properly, astronauts started running on the COLBERT and pronounced it a big improvement on the previous treadmill.

Countermeasures to bone demineralization 85

Figure 5.4 After more than four hours of running in place, NASA astronaut Sunita Williams completed the 2007 Boston Marathon while orbiting Earth aboard the International Space Station (ISS). Williams began running the marathon at 10.00 am EDT as the race kicked off in Boston and the ISS circled Earth at 28,163 km/hr. At about 2.24 pm EDT, she radioed Mission Control that she'd completed the race with an unofficial time of about 4 hr 24 min, marking the first time an entrant had competed from orbit. The Boston Athletic Association issued her the bib number 14,000, which Williams taped to the front of her treadmill during the event. On Earth, Williams' fellow NASA astronaut Karen Nyberg completed the Boston Marathon in 3 hr 32 min. Williams ran the Boston Marathon as many of her fellow crewmembers slept, although Expedition 14 commander Michael Lopez-Alegria and Expedition 15 flight engineer Oleg Kotov prepared drink pouches and orange slices for her during the race. Image courtesy: NASA.

86 Bone loss

ISS astronauts, ECM crewmembers, in an effort to maintain their bone density on long missions, will train like marathon runners. For example, Sunita Williams (Figure 5.4) became the first astronaut to run a marathon in orbit when she ran the distance in conjunction with the Boston Marathon in 2007. Initially, Mission Control wasn't too enthused with Williams' idea because the start of the race conflicted with the crew's sleep cycle and they were worried that all the pounding on the treadmill would wake sleeping crewmembers. Fortunately, her fellow crewmembers were flexible and gave her the green light.

Despite several hours spent exercising every day, astronauts – even marathon runners like Williams – will still return to Earth with reduced bone density, although there are large differences between individuals: at one extreme, there is the case of NASA astronaut, David Wolf, who spent 128 days onboard Mir and lost up to 12% of his bone mass in certain areas, and at the other extreme is the case of cosmonaut Yuri Romanenko, who spent 326 days in space (MIR EO-3) but didn't show any significant bone loss (in fact, Romanenko stood up unaided following the landing and ran 100 m the next day!). But the cases of Wolf and Romanenko are the exception. Despite four decades of investigating ways to prevent bone loss, no astronaut has ever gone into space and not experienced a certain degree of bone loss. It's a problem characterized by all sorts of idiosyncrasies. While drug intervention, nutritional manipulation, and exercise help, none of these strategies alone represents a solution to the problem. Perhaps this is because scientists are approaching the problem from the wrong angle. Rather than treating the effects of zero gravity system by system, why not treat the root cause of zero gravity by simply restoring gravity?

ARTIFICIAL GRAVITY

Artificial gravity (Figure 5.5) is a common technology in science fiction movies. For example, in the *Star Trek* universe, artificial gravity is achieved by the use of "gravity plating" embedded in a starship's deck. In Gene Roddenberry's *Andromeda*, set thousands of years in the future, gravity field generators not only provide artificial gravity for the crew, but also reduce the inertial mass of ships to less than a kilogram. This weight reduction increases the efficiency of the ship's magneto-plasma dynamic drive, which allows them to go from a stop to percentages of light-speed very quickly. Computer games have also used artificial gravity as a setting. For example, the classic computer game *Halo: Combat Evolved* is set on an artificial ringworld that creates artificial gravity by computer-controlled rotational spin (inspired by Larry Niven's *Ringworld* that featured a habitat that created artificial gravity through rotation). There are several other imaginative examples of using artificial gravity in the world of movies and computer games, but perhaps the most vivid and most striking is the science fiction classic *2001: A Space Odyssey*, which features a rotating centrifuge in the *Discovery* spacecraft.

While astronauts would love to have an *Andromeda* gravity-field generator, it is likely this technology won't be available for some time, whereas the technology depicted in Stanley Kubrick's classic is much closer to the horizon. In fact, scientists

Artificial gravity 87

Figure 5.5 Artificial gravity. This is a 1969 station concept, designed to be assembled on orbit from spent Apollo stages. The station was to rotate on its central axis to produce artificial gravity. Image courtesy: NASA.

working for NASA's Artificial Gravity program are using a short-radius centrifuge (SRC) built by Wyle Laboratories to research ways of reducing the effect of zero gravity on astronauts' bodies. Known as the "Big AG", NASA's artificial gravity research uses a SRC (Figure 5.6) capable of spinning two riders simultaneously. As you can see in the photo, the SRC has two arms that extend in opposite directions from a central pivot point. As the arms swing on the pivot, centrifugal forces create G-loads along the rider's body axis (head to feet) proportional to the rate of rotation. A device mounted on the foot-plate measures the G-forces at the feet, while other biomedical instrumentation mounted on the arms records heart rate and other physiological parameters.

For most of the tests, NASA spins the SRC at 17.3 revolutions per minute (rpm). At this speed, riders feel their feet pressing against the foot plate a little more firmly and notice some mild turbulence as air flows past the arm's windshield. When spinning in the dark, the SRC is so quiet that riders have very few clues that they're moving at all. It may sound as if NASA is on its way to solving the problem of artificial gravity, but like any new technology, there are problems. One snag is the effect of rotation on the astronauts as they spin. Spinning at 17.3 rpm on NASA's

88 Bone loss

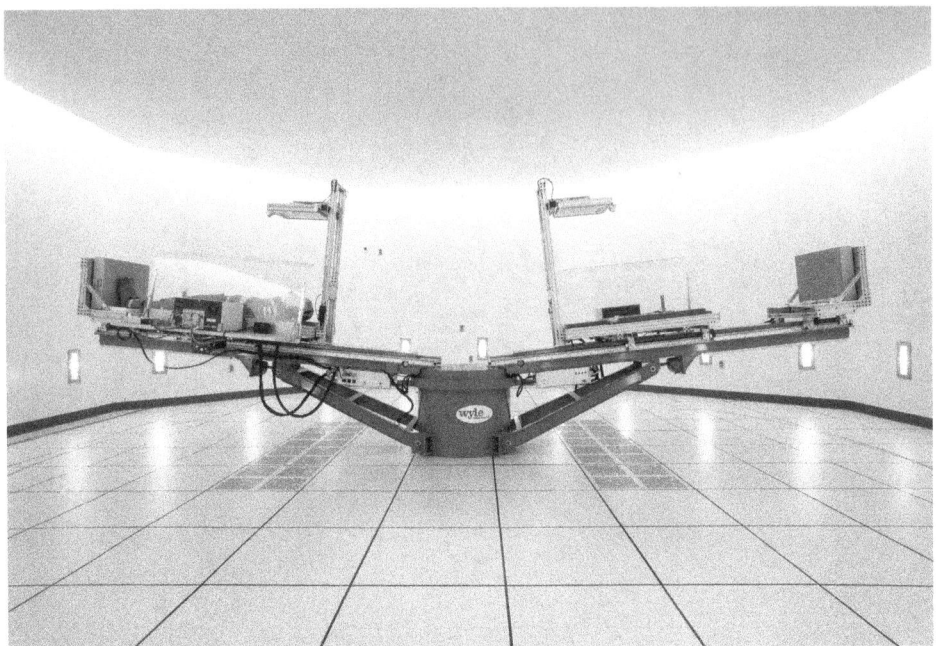

Figure 5.6 NASA's artificial gravity program uses a short-radius centrifuge built by Wyle Laboratories to research ways of reducing the effect of zero gravity on astronauts. Image courtesy: NASA.

SRC is fine if you're spinning for a few minutes, but after a while, some riders feel dizzy, nauseous, and disoriented. Of course, the solution is simply to reduce the spin rate, because it's known that a spin rate of 2 rpm or less produces no adverse effects, but the problem with such a low spin rate is that it doesn't produce sufficient gravity! It's a classic Catch-22, but even if the NASA scientists solve it, there's still the problem of angular movement to overcome. That's because high angular velocities produce high levels of Coriolis forces – angular moments (the amount of energy to spin) that would require a propulsion system of some kind to spin up (or spin down). Also, if parts of the spaceship are intentionally not spinning, friction and torque will cause the rate of spin to decrease as well as causing the otherwise stationary parts to spin. To compensate for these effects, fly wheels and thrusters would be needed to keep the appropriate sections of a spacecraft spinning or not.

The problems of artificial gravity haven't stopped mission planners from including it in future interplanetary mission architectures. For example, the Human Outer Planets Exploration (HOPE) mission (see Chapter 2) conceived by NASA's Institute for Advanced Concepts (NIAC) in 2003 integrated an artificial gravity capability onboard their bi-modal nuclear thermal rocket (BNTR) vehicle designed to ensure crew health and fitness on a long-duration mission to Callisto that was projected to last five years. In NIAC's mission design, the crew transfer vehicle initiates vehicle rotation at about 4 rpm to provide the crew with a Mars gravity environment (0.38 times Earth gravity) during the outbound transit. A higher rotation rate of ~6 rpm would provide about 0.8 G during the return leg of the mission to help re-acclimate the crew to Earth's gravity.

Unfortunately, artificial gravity research has suffered recently (artificial gravity studies at Johnson Space Center were shut down several years ago) because NASA was spending all its money returning astronauts to the Moon. But, with the Moon out of the picture, thanks to NASA's new direction, there could be a resurgence in artificial gravity work. The first step may be to place a centrifuge on the ISS to help answer the question of artificial gravity's effectiveness in space. Operation of an ISS centrifuge[1] would also answer questions such as how fast astronauts should be spun and whether crewmembers encounter problems with dizziness or vertigo. If artificial gravity is proved effective, spacecraft designers will get to work deciding how best to incorporate artificial gravity into an interplanetary mission. While one option would be to install a centrifuge inside the vehicle, another option would be to simply rotate the spacecraft. To see how this would be done, we'll take a look at some work being done by NASA engineer, Kent Joosten.

Joosten's spinning spaceships

Joosten's idea was to design a spacecraft that would spin at a rate of 4 rpm. The spin rate was based on tests conducted at the Pensacola Slowly Rotating Room in the 1960s and 1970s. During these tests, it was determined that at a speed of 4 rpm, some individuals would suffer no motion sickness, while others would adapt within a few days. Using a spacecraft (Figure 5.7) with a rotational radius of 56 m, Joosten calculated that such a configuration would produce 1 G at 4 rpm.

As you can see in the diagram, Joosten's spacecraft is an axis-spinner. The control jets located at the end of the arm mean they possess large moment arms, which, in turn, mean that using a moderate level of thrust, the configuration would be spinning at the desired 4 rpm within two days. This spin rate would obviously generate high centripetal tension loads, so Joosten decided to use ultra-high modulus graphite for the spars due to the material's extreme stiffness. For the crew habitat, Joosten chose an inflatable structure attached to the end of one of the arms. Compared to spacecraft to date, the complexity of Joosten's spacecraft would be significantly reduced because there would be no need for microgravity systems. This would mean the waste disposal system, hygiene systems, and sinks wouldn't need vacuums to control free-floating debris as is the case onboard the ISS today. An additional feature of the 1-G crew habitat would be the inclusion of Earth-based

---

[1] NASA had plans to include a centrifuge onboard the ISS. The Centrifuge Accommodations Module (CAM) would have been attached to the Harmony module of the ISS and would have exposed biological specimens to artificial gravity levels of between 0.01 and 2 G. It was cancelled in 2005 because of ISS cost overruns and scheduling problems in Shuttle assembly flights. A partly built shell of the CAM is on display in an outdoor exhibit at the Tsukuba Space Center in Japan.

90 Bone loss

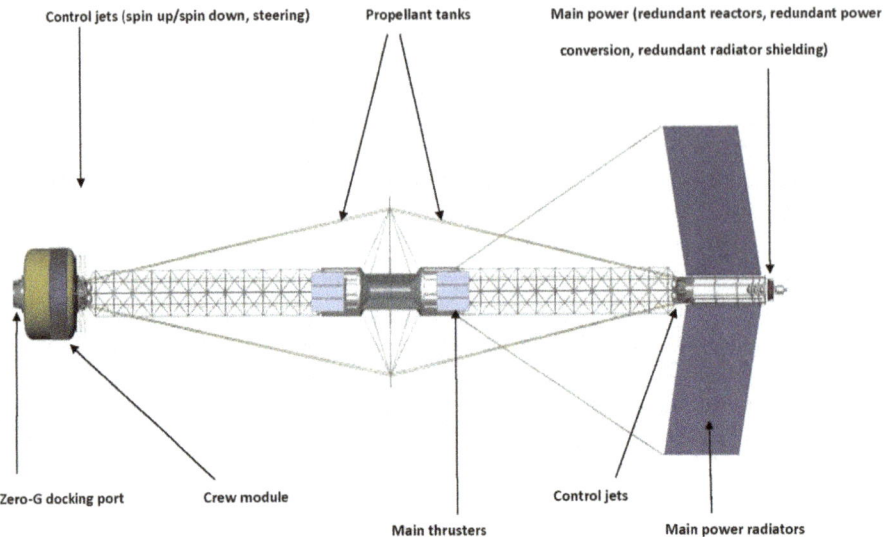

Figure 5.7 Promising research is being conducted in the area of artificial gravity generated by rotation of the entire space vehicle, as shown in this image. Artificial gravity is the centripetal force generated in a rotating vehicle and is proportional to the product of the square of angular velocity and the radius of rotation. For a particular G-level, there is a trade-off between velocity of rotation and radius and, since increased radius is more expensive to achieve than velocity, most research is directed at defining the highest rotation rate to which humans can adapt. Image courtesy: NASA.

comfort items such as chairs. No doubt, the habitat would include exercise devices, but these would be included more for crew relaxation than as a means to maintain bone density.

If artificial gravity is proven effective at preventing bone loss and if engineers can make the system work without making the crew sick or dizzy, it's possible that one day, ECM astronauts will travel to the outer planets on spinning spaceships similar to the one designed by Joosten. If it works, artificial gravity will be an elegant solution to the problem of bone loss. If it doesn't work, then perhaps the answer will be a combination of the medical countermeasures described here and artificial gravity – a solution that would represent a combination of clever medicine and clever spacecraft.

6

Behavior and performance

"Nobody in the history of mankind has ever experienced the Earth as a pale, insignificant blue dot in the sky. What that might do to a crew member, nobody knows."

Dr Nick Kanas, NASA psychologist

"When early explorers left their home countries on the seas, they didn't see their home countries anymore. They didn't even have a dot to look at. It was out of sight on the other side of the world. It is not like we are reinventing the wheel. We are just doing the same thing in a different environment that was just as demanding then."

Walter Sipes, NASA psychologist, Johnson Space Center, Houston

If you take an interest in the future of manned spaceflight, you will no doubt have read articles about the list of problems an interplanetary journey would entail. Somewhere on that list, you will probably have read about the psychological challenges of sending astronauts on multi-year voyages. Some scientists are so worried about what might happen that they've set up simulated Mars missions to test the psychological effect of such a trip. Recently, the European Space Agency (ESA) locked six "astronauts" in a tube (Figure 6.1) for months. For the volunteers who emerged from their 105-day test inside a titanium chrysalis, it seems like it wasn't a day to soon. While the volunteers claimed they managed to keep up a fairly good *esprit de corps*, they also reported mind-numbing boredom that apparently damaged their capacity to learn and tampered with their ability to focus. Some tried to pass the time by playing cards and dice, and by attempting to learn Russian, but the brain-deadening boredom of the endless days and featureless enclosure made it impossible for some to remember any of the Russian words studied during the simulation. And then there was the issue of interpersonal tensions. A couple of crewmembers nearly came to blows over the use of the treadmill in the module's tiny gym, while another didn't enjoy eating space rations selected by others.

It would appear that the experience of the 105-day volunteers doesn't bode well for the actual astronauts who will face not only boredom, but a myriad other dangers ranging from rapid decompression to radiation sickness. Needless to say, the

92 Behavior and performance

Figure 6.1 The MARS500 test crew passes the halfway point in their 105-day Mars mission simulation. The crew included two European members selected by the European Space Agency: Oliver Knickel, a mechanical engineer in the German army, and Cyrille Fournier, an airline pilot from France. The remaining four were Russians: cosmonauts Sergei Ryazansky (Commander) and Oleg Artemyev, Alexei Baranov, a medical doctor, and Alexei Shpakov, a sports physiologist. They lived in the specially designed isolation facility in Moscow for 105 days, beginning on March 31st, 2009. Their tasks were similar to those they would have had on a real space mission. They dealt with simulated emergencies and coped with an operative communication delay of up to 20 min each way. The initial 105-day study was the precursor to a complete simulation of a fully fledged 500-day mission to Mars that started in 2010. Image courtesy: European Space Agency.

experiences of the volunteers provided the naysayer psychologists with more ammunition. Astronauts will go crazy, they said. How can we send astronauts on a 1,000-day trip into space when they can't even survive 100? The response from the scientists was predictable. The simulation provided invaluable data for the real thing, they replied, sticking to their guns. In fact, they already had plans for a longer mission. In June 2010, another six volunteers were locked inside a similar mockup (Figure 6.2) in an experiment that is simulating (it's probably still going on as you read this) an entire 520-day mission to Mars. The 105-day guys got off easy!

The mission, dubbed MARS500, may sound like a sadistic reality TV show concocted by Mark Burnett, creator of *Survivor*: seal six men inside a claustrophobic mock spaceship, limit their contact with the outside world, and keep them there for nearly a year and a half. While MARS500 isn't a TV show, cameras are rolling,

Figure 6.2 The MARS500 facility. Image courtesy: European Space Agency.

recording all that happens inside the "spaceship". Like the 105-day stay, MARS500 is a research project to find out how well a crew can endure the journey to the Red Planet. The mock spaceship comprises a series of interconnected steel canisters, with a total volume of some 550 cubic meters. During their "journey", the volunteers have to survive on limited food rations and their only communication with the outside world is via email, delayed and occasionally disrupted to imitate the effects of space travel. The crewmembers' days are divided into eight-hour chunks comprising sleep, work, and leisure. They have two days off in a week, except during simulated emergencies, of which several are planned. To pass the time, the crew play video games as part of ESA's project to develop personalized software to interact with crews on future space missions. The highlight of their "voyage" will be a simulated spacewalk on Mars, which will take place in a large sandpit. Ahead of the start of the mission, psychologists warned that months of simulated space travel would push the team to the limits of endurance as they grew increasingly tired of each other. Such statements were based on previous experience; an earlier eight-month simulation study carried out at the same institute between 1999 and 2000 ran into trouble when a female Canadian scientist complained she had been forcibly kissed by the Russian team captain and when two Russian crewmembers had a fist fight!

Just like the 105-day venture, scientists argued MARS500 would yield valuable data, and they're looking forward to seeing what will happen inside the hangar-based spaceship. In reality, despite all the justifications of the scientists, MARS500 will amount to little more than observing half a dozen crewmembers sitting inside a tin can with no sun, no fresh water, and no alcohol. While the intent is to simulate a mission to Mars, the whole exercise is more *Red Dwarf* than Red Planet because for mission planners anxious about the effects of tedium and lethargy on Mars crews, there is already an abundance of knowledge gleaned from polar expeditions. Rather than spending millions of dollars building a tin can in a Moscow hangar, all these scientists had to do was wander down to their local library and read accounts of those who led men on expeditions far more arduous than any trip to Mars. You see, human intuition tells us that the experience of an ECM will be so different from any mission that has preceded it that unearthly changes will manifest themselves in the crew, hence the need for extensive research. Psychologists search for analogs for the space environment and see in those analogs only examples of human frailty; if only they'd bothered to read a few books once in a while, they would see that in the course of exploration, there are many examples of human greatness and history provides us with many examples of how people have performed admirably under stressful circumstances. Perhaps one of the best examples of how a crew coped with the problems of a multi-year mission is the case of Ernest Shackleton's Imperial Trans-Antarctic Expedition in 1914.

SHACKLETON

Shackleton's objective was to cross the Antarctic continent, but when he and his crew were still hundreds of kilometers from the intended base, his ship, the *Endurance* (Figure 6.3), first became trapped and was then crushed by pack ice. Shackleton and his crew abandoned the *Endurance* before she sank. He and his crew survived for two months on an ice floe, hoping it would drift to an island where stores were cached. After the floe broke in two, Shackleton ordered the crew into the lifeboats and headed for Elephant Island (Figure 6.4) – a place not much more hospitable than the surface of Mars. Knowing Elephant Island was far from the shipping lanes, Shackleton risked an open-boat journey to a distant South Georgia whaling station, where help was available. Taking the strongest lifeboat, the *James Caird*, Shackleton chose five crewmembers, including Frank Worsley, *Endurance*'s captain. For 15 days, the *James Caird* sailed through some of the roughest waters in the world, at the mercy of hurricane-force winds and mountainous seas. Thanks to Worsley's extraordinary navigating skills (all without the aid of a global positioning system or weather radar!), the *Caird* reached South Georgia and after a 36-hr crossing of South Georgia's mountains, the party arrived at the whaling station at Stromness.

Shackleton immediately set to work organizing the rescue of the rest of his crew on Elephant Island. His first three attempts failed due to sea ice, but with the help of the Chilean government's sea tug, the *Yelcho*, Shackleton successfully rescued the remainder of this crew. While Shackleton's expedition is perhaps the most famous polar exploit, there were several other explorers who led men into inhospitable terrain and returned safely. Take Norwegian explorer Fridtjof Nansen's *Fram* expedition, for example.

Figure 6.3 Shackleton's ship, the *Endurance*, trapped in the ice. Photo taken by Frank Worsley. Image courtesy: Wikimedia.

96 Behavior and performance

Figure 6.4 Launching the *James Caird* from the shore of Elephant Island, April 24th, 1916. Image courtesy: Wikimedia.

NANSEN

The *Fram* expedition was an attempt by Nansen (Figure 6.5) to reach the geographical North Pole by harnessing the natural east–west current of the Arctic Ocean. Nansen's plan was simply to beach the *Fram* (Figure 6.6) on the pack ice and hope the movement of the ice would transport the ship and its crew towards the North Pole. To do this, he supervised the construction of the *Fram*, a custom-built vessel with a rounded hull, designed to withstand prolonged pressure from the ice. Departing in June 1893, the *Fram* sailed across the Barents Sea towards Novaya Zemlya and then to the North Russian settlement of Khabarova before traversing the Kara Sea, an expanse of water for which charts were incomplete. By September, Nansen's progress had been hampered by heavy ice and he set course for the New Siberian Islands, where he hoped to find open water, before entering the pack ice at 80° latitude. In October, the *Fram* was moored on the ice in preparation for winter. Unfortunately, the movement of the ice was unpredictable and after a few weeks, the *Fram* was actually south of its original position. In common with the volunteers who participated in the Mars simulation experiment, Nansen's crew experienced boredom and frustration. Fights broke out and men became irritated at the lack of inactivity. Unlike the crew of those in the Mars simulators, Nansen's crew had nowhere to go, but they coped very well nonetheless. In March 1894, the *Fram* was still held in the ice and progress continued to be slow – so slow that Nansen came up with another

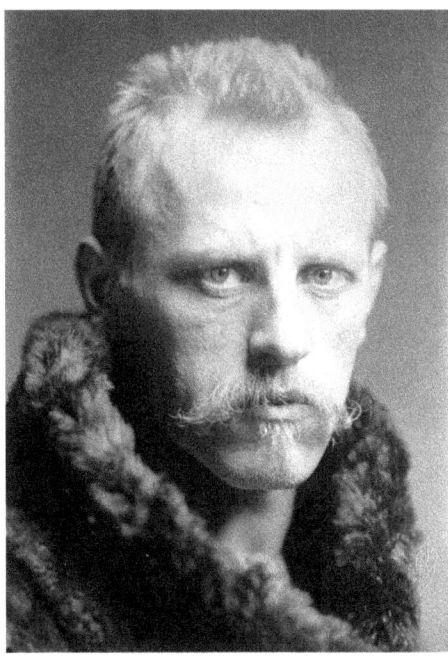

Figure 6.5 Fridtjof Nansen. Image courtesy: Wikimedia.

plan to reach the pole. In September 1894, Nansen decided that he and one crewmember would leave the ship and ski to the pole. After reaching the pole, the pair would continue on to Franz Josef Land and then to Spitzbergen, where they planned on hitching a ride home to Norway.

By the time Nansen and his companion, Hjalmar Johansen, finally departed on their quest to reach the pole on March 14th, 1895, the crew had been away from home for almost two years. Remember, this was *1895*, not 1995. The crew didn't have MP3 players or access to Mission Control and there was no such thing as Facebook or texting. In fact, there was no communication at all. And yet, despite their deprivations and the brutal cold, the crew functioned very well. Behavioral problems? Not these guys. Psychiatric disorders? Are you kidding? And let's not forget who these guys were. They weren't highly qualified astronauts with Ph.D.s and thousands of hours of mission-specific training under their belts. With the exception of the ship's doctor, a couple of engineers, and a Reserve Army lieutenant, the crew – just like the crew of the *Endurance* – were sailors with little, if any, formal education. They almost certainly would have made good astronauts if they had been born 100 years later! Anyway, let's get back to the expedition. Before leaving his crew, Nansen appointed the expedition's second-in-command, Otto Sverdrup, as leader, tasking him to continue with the drift towards the Atlantic Ocean unless circumstances dictated abandoning the ship. Nansen and Johansen had allowed themselves 50 days to cover the 660 km to the pole. On April 3rd, after weeks of travel in temperatures of 40°C below and failing spirits, the pair finally reached their most northerly latitude of 86° 13.6′N, almost three degrees beyond the previous farthest north mark. Ahead of them was a chaotic expanse of ice-blocks that thwarted any attempt to continue, prompting Nansen to decide it was time to retreat to Franz Josef Land. Killing their sled-dogs at regular intervals to eat along the way, the pair finally sighted land at the end of July. They continued their progress, but by the end of August, with winter looming, Nansen decided to make preparations for their third winter in the Arctic. Their base for their winter quarters was a primitive hut (Figure 6.7), which they improvised by stretching walrus skins over the top to form a roof. They subsisted on a diet of bear, walrus, and seal, and entertained themselves by reading Nansen's sailing almanac by the light of a blubber lamp. In May, 1896, after spending the best

98 Behavior and performance

Figure 6.6 Nansen's ship, the *Fram*, pictured in the Fram Museum. Image courtesy: Frammuseet, Oslo.

Figure 6.7 The hut that Nansen and Johansen used as their winter quarters. Image courtesy: Wikimedia.

part of nine months in the hut, Nansen and Johansen set forth again, and after a month of kayak travel, arrived at Cape Flora, where they met Fredrick Jackson, who had organized his own expedition to Franz Josef Land after being rejected for Nansen's *Fram* venture. Jackson's supply ship *Windward* was due to call that summer, and Nansen and Johansen simply had to wait a few more weeks before being rescued. In August, the *Windward* departed with Nansen and Johansen onboard and after a few days at sea, the pair arrived back in the Norwegian port of Vardø. Two weeks after their arrival, they were joined by Sverdrup, who arrived in Vardø with the *Fram* and the rest of the crew, none of whom reported any behavioral symptoms or psychological problems from having been away from home for more than three years or from having suffered the hardships of having spent so long in one of the most inhospitable environments on Earth.

Of course, Shackleton and Nansen aren't the only explorers who ventured into the unknown on multi-year expeditions. The history of polar exploration is rife with tales of such heroism, self-sacrifice, and conquest, and thanks to meticulous diaries maintained by these explorers, today's mission planners have an excellent understanding of the demands future interplanetary astronauts will face. For some reason, this wealth of information seems to have been ignored by those concocting analog missions such as the MARS500 boondoggle, yet the insight gleaned from an understanding of Shackleton and co.'s exploits is far more helpful than data gleaned from these psycho-social isolation experiments. Why? Well, let's face it, no research ethics board is ever going to approve a confinement experiment that is conducted under the rule "Once you're in, you cannot get out until the study is finished, no matter what happens". You see, those poor isolation lab rats stuck inside that tin can near Moscow Airport (where the MARS500 experiment is taking place) know that in a real emergency, evacuation and rescue will be at hand. ECM crews will know the opposite. So where's the efficacy or relevance of the study? Enough said. Having said that, an interplanetary trip will impose some unique stressors (polar explorers didn't have to deal with radiation or loss of visual contact with the Earth, for example) upon the crew, and while it is certainly not worthwhile spending money on any more MARS500 simulations, it's still necessary to discuss how future crews may be affected.

INTERPLANETARY STRESSORS

In common with Shackleton's and Nansen's crews, those embarking on an interplanetary journey will be subject to isolation and confinement along with a host of other behavioral stressors (Table 6.1). First, during the transit to and from the planet, crewmembers will be physically isolated from Earth, while on the surface, they will be confined to a habitat no larger than a school bus, surrounded by the most hostile environment faced by humans to date. Despite rigorous selection procedures, it's only natural that crewmembers separated from family and friends will feel degrees of emotional deprivation. Social confinement will be exacerbated by absence of privacy, and little separation will exist between work and leisure because the living and working spaces will be so close to one another and each crewmember will interact with the same group of individuals in both types of activities.

> "All the necessary conditions to perpetrate a murder are met by locking two men in a cabin of 18 by 20 feet for two months."
>
> Cosmonaut, Valery Ryumin

Table 6.1. Behavioral stressors of long-duration spaceflight.

Psychological	Psycho-social	Human factors	Habitability
Isolation and confinement	Team coordination demands	High and low workloads	Limited hygiene
Limited abort options	Interpersonal tension	Limited communication	Chronic noise exposure
High-risk condition	Family life disruption	Limited equipment	Limited sleep facilities
Mission complexity	Enforced interpersonal contact	Mission danger, risk of equipment malfunction	Lighting and illumination
Hostile environment	Crew factors (size, gender)	Adaptation to environment	Lack of privacy
Sensory stimuli altered	Multicultural issues	Food restrictions	Isolation from support

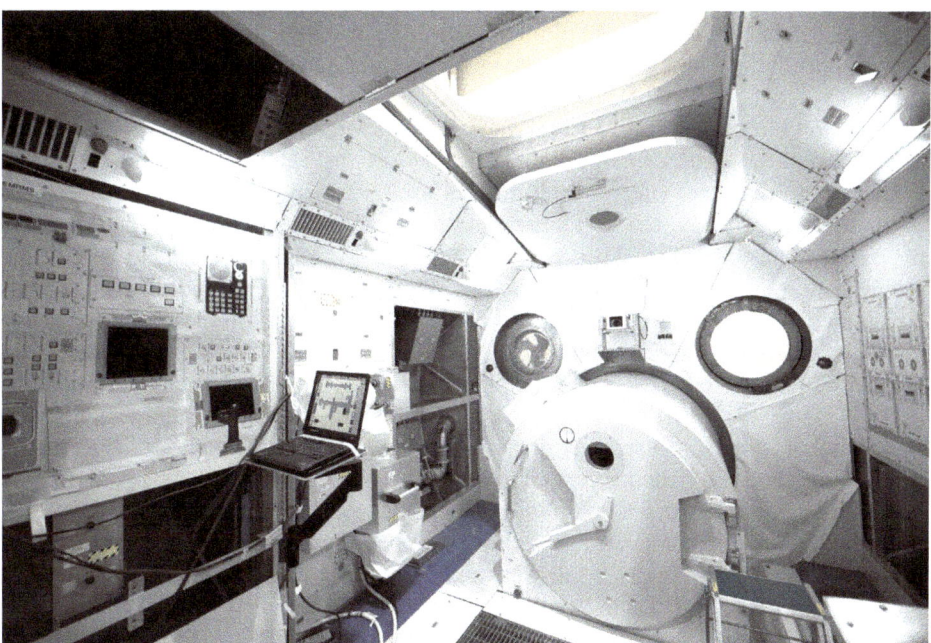

Figure 6.8 Living space during exploration class missions will be cramped. Image courtesy: NASA.

As Ryumin's quote suggests, sustained, close personal contact with other crewmembers will be extremely stressful – a situation compounded by isolation, dangers posed by radiation, equipment malfunctions, and boredom. Polar explorers faced with interpersonal problems had the option of removing themselves temporarily from the source of stress by simply going for a walk. Unfortunately, interplanetary explorers, confined to a living space (Figure 6.8) the size of a small motor-home, shared with three to five other crewmembers, will not have this option. Inevitably, during the cruise phase of the voyage, each crewmember's repertoire of jokes, personal experiences, and anecdotes will become increasingly familiar. Mannerisms, which were initially innocuous, will be exaggerated in the minds of a crew subject to the constraints of their extended confinement. Eventually, the most minor irritation will assume unreasonable proportions and will force some crewmembers to retreat. Unfortunately, the only escape will be the lavatory or the small coffin-sized compartment that serves as a sleep cubicle. But don't feel too bad for the astronauts. Remember Nansen and Johansen in their hut? Take a look at that picture again, compare those living conditions with the habitat portrayed in Figure 6.8, and then ask yourself which one you would rather spend three years in!

Okay, so isolation and claustrophobia will be a problem, but thanks to the experiences of Shackleton and Nansen, we know these are survivable problems. But what about interpersonal problems? Surely, mission planners can't expect a crew to survive three or more years without someone venting their frustrations? Well, if we examine the diaries of polar explorers and, more recently, those of personnel who have spent winters in the Antarctic, we get some idea of the sort of interpersonal problems astronauts might face. For example, it is very likely that the phenomenon of *psychological closing* will be observed among crewmembers. Psychological closing is manifested by decreased communication intensity as well as increased filtration of the scope and content of crew communication. All it means is that crewmembers tend to conceal medical and psychological problems and also demonstrate preferences in contacts with certain Mission Control personnel. A related phenomenon is *autonomization*, which is expressed by crew egocentrism and is considered a natural and even necessary stage of the formation of a cohesive group of isolated and confined individuals. The problem with this phenomenon occurs when it is manifested by Mission Control personnel being perceived as opponents and not partners. In such a situation, the crew becomes critical in their discussions with Mission Control, resulting in the compromise of operational effectiveness. A good example of the phenomenon of autonomization was the crew of Skylab-4, who complained to Mission Control about being overworked. Closely related to autonomization is the phenomenon of *displacement*, which occurs when crewmembers experience high levels of anxiety and interpersonal conflict that cannot be resolved directly. In such a situation, the unpleasant effects may sometimes be externalized to Mission Control personnel. The cause of the displacement is a coping strategy, allowing crewmembers to avoid open conflict by venting their frustrations on the unfortunate Mission Control personnel. Although the strategy works for a while, in the long term, it encourages negative feelings, territorial behavior, and a disintegration of group cohesion.

Psychological closing, autonomization, displacement, and a host of other psychological problems have been observed in environments similar to spaceflight. For example, consider nuclear submarine crews who stay underwater for up to six months. Since nuclear submariners are among the most thoroughly screened individuals (next to astronauts) on Earth, it's not surprising that the incidence of psychiatric illnesses among crewmembers is relatively low compared to the general population, but even this group of undersea workers experience anxiety, depression, and interpersonal problems. These similarities between groups have prompted scientists to examine other analog environments to help them understand what sort of behavioral problems interplanetary astronauts might face. For example, while the nuclear submarine environment replicates to some degree the confines of a spacecraft, many researchers agree the most useful analog is the Antarctic research station (Figure 6.9). This is due to the extreme environment, the extended tours of duty, and the limited contact with the outside world. Other features Antarctic research stations share with orbiting spacecraft include heterogeneity of crewmembers, high skill levels, organizational similarities, and the rotational structure of tours of duty. In fact, the conditions in the Antarctic and onboard orbiting space stations are considered so similar (Table 6.2) that Antarctica research stations such as Concordia have served as a primary source of psycho-sociological data for predicting behavior onboard future interplanetary missions. While analogs are not perfect simulations of spaceflight (crew characteristics, screening procedures, mission

Figure 6.9 Concordia is one of the most recent and most modern research stations in Antarctica, and one of only three on the continent. The station was built and is operated by French Polar Institute and the Italian Antarctic Program. The remote outpost, located at an altitude of 3,200 m, has been permanently manned since 2005. Image courtesy: G. Dargaud and Wikimedia.

objectives, and duration are different for each analog), the environments are the only tool researchers have to study the behavioral impacts of confinement, isolation, and prolonged periods of stress.

In addition to the use of analogs, scientists have used ground-based space simulators such as the MARS500 tin can. For example, in 1990, ESA locked six civilians with science and engineering backgrounds inside a hyperbaric chamber for four weeks as part of a simulation experiment designed to replicate the daily activities of astronauts. The Isolation Study for European Manned Space Infrastructure (ISEMSI), as it was known, subjected the volunteers to various psycho-social tests designed to identify social and emotional conflicts and to record and analyze patterns of communication between crewmembers. The study observed various changes in interpersonal and adjustment reactions but on the whole, nothing unusual was observed. A less successful study, which resulted in fist fights and charges of sexual harassment, took place during the Simulation of Flight of International Crew on Space Station (SFINCSS) experiment conducted by the Russian Institute for the Study of Biomedical Problems in 1999 – the same place

Table 6.2. Comparison of psychologically relevant factors.

	Orbital ISS mission	Winter-over in Antarctica	Mars mission	Polar exploration
Duration (months)	4–6	9–12	16–36	12–48
Distance to Earth (km)	300–400	N/A	60–400 million	N/A
Crew size	3–6	15–100	4–6	>4
Degree of isolation	Low to high	Medium	Extremely high	Extremely high
Crew autonomy	Low	High	Extremely high	Extremely high
Evacuation in case of emergency	Yes	No	No	No
In-flight support measures				
Outside monitoring	Yes	Yes	Yes	No
Two-way communication	Yes	Yes	Restricted	No
Email/up/down-link	Yes	Yes	Yes	No
Internet access	Yes	Yes	No	No
Re-supply	Yes	No	No	No
Visitors	Yes	No	No	No
Visual link to Earth	Yes	Yes	No	N/A

where the MARS500 tin can experiment is taking place! During a party celebrating New Year's Eve inside a Mir simulator, Canadian Judith Lapierre was pulled out of view of the module's observation cameras and forcibly kissed by a Russian crewmember. The study had been designed to evaluate and observe social interactions of mixed-gender crew in which crewmembers were housed in separate modules and executed different flight programs. Until the kissing incident, the study had largely gone unreported by the media. Following the fist fight and sexual harassment incident, conditions deteriorated to the point where the commander requested two crewmembers either be withdrawn from the study or have a hatch closed between two chambers to prevent further interaction – presumably, if it had been a real mission, the commander would have flushed them through the airlock! Ultimately, the failure of the SFINCSS was testament to just how soft Western society has become; this supposedly highly trained group had all the creature comforts that were denied Shackleton and Nansen, yet they couldn't even make it through 110 days. Perhaps the crew should have read Shackleton's book before entering the simulator!

So far, we've discussed isolation, interpersonal conflict, and withdrawal, none of which has the potential to seriously compromise the mission. Sure, crewmembers might feel a little depressed once in a while. They may even feel like strangling their crewmates after they've told the same joke once too often. But compared to the rigors of being stranded on a windswept island eating penguin meat (Figure 6.10), life won't be that bad. But what about boredom? Surely *that* will cause problems for the crew. After all, there are only so many films you can watch and only so many books you can read.

Interplanetary stressors 105

Figure 6.10 Shackleton's party, left behind on Elephant Island, used the two remaining lifeboats to make a hut. A blubber stove provided heat and was used as a cooker to prepare penguin meat. Image courtesy: National Library of Australia.

"Oh! At times this inactivity crushes one's very soul; one's life seems as dark as the winter night outside ... I feel I must break through this deadness, this inertia, and find some outlet for my energies. Can't something happen?"
Dr Fridtjof Nansen, Legendary Norwegian Polar Explorer, 1897

Well, unlike the other stressors we've discussed, which were common to polar exploration and interplanetary travel, the issue of boredom is probably the least cause for concern for mission psychologists. Between the high-tempo flurries of activity following departure from Earth orbit and preparation for arrival at the planet, crewmembers will be required to perform all sorts of tasks ranging from scheduled systems checks and emergency abort rehearsals to laboratory work and equipment repairs. Forget all those psychological studies of bored navy personnel at isolated postings pining for the bright lights of San Diego. Interplanetary crewmembers will be busy. Real busy. In fact, their main problem won't be boredom, but overwork. I have friends who regularly spend time in Mars analogs in the Arctic who've witnessed commanders' regularly ordering their teams to stop work at 10 pm. Rather than worry about boredom, psychologists should worry more about avoiding the debacle of the 1973 Skylab-4 mission (Figure 6.11) described briefly in Chapter 1. Skylab-4 comprised entirely rookie astronauts who complained they had too much work. Unsurprisingly, because the Skylab-4 astronauts had been

106 Behavior and performance

Figure 6.11 The Skylab-4 crew. Left to right: Commander Gerald Carr, Science Pilot Edward Gibson, and Pilot William Pogue. Image courtesy: NASA.

given a similar workload to previous Skylab crews, Mission Control was unsympathetic to the complaints of Commander Gerald Carr, Pilot William Pogue, and Science Pilot Edward Gibson, and told the crew to continue with the assigned work. The astronauts, unhappy with the attitude of Mission Control, switched off the radio and staged a mutiny by declaring an unscheduled rest day. This act of insubordination resulted in a NASA rule stating that at least one International Space Station (ISS) crewmember must now have previous spaceflight experience.

So, still no cause for concern. But what about possible meltdowns? After all, despite rigorous selection criteria, psychiatric and psychological events have occurred during space missions. For example, there is anecdotal evidence the Russians once launched a rescue mission to the Mir space station for the purpose of returning one stress-stricken cosmonaut to Earth. Psychologists argue that even though astronauts represent the most highly trained and motivated workers on and off the Earth, the fact remains that even this select group has a threshold beyond which things may get out of hand. They may have a point. After all, interplanetary crewmembers will be exposed to a much higher level of autonomy and long-term confinement and isolation than any previous space crew. Some psychologists have suggested this may increase the psychological risks related to individual crewmember performance and will produce new psychological challenges never before experienced. For example, *asthenia* is a common psychiatric condition that may affect astronauts. The syndrome is defined as a weakness of the nervous system, resulting in fatigue, irritability, concentration difficulties, restlessness, and physical weakness. In the Russian space program, *asthenization* is carefully monitored and countermeasures are employed to prevent it, while in the United States, the syndrome is addressed from a different perspective, resulting in NASA flight surgeons possibly under-detecting the syndrome. While asthenia might be relatively easy to treat, what happens if a crewmember becomes psychotic or violent towards other crewmembers? It's a scenario that already has deep resonance in popular culture; Christian Alvart's *Pandorum*, Duncan Jones's award-winning *Moon*, and Andrei Tarkovsky's *Solaris* are science fiction movies that employ the device of extended-duration space travel to explore themes of crewmembers becoming unhinged. No doubt, at least two crewmembers will be trained in counseling, crisis intervention techniques, and the administration of psychoactive medications, but how will the belligerent crewmember be restrained? Will future interplanetary spacecraft carry straightjackets? The answer is probably yes, but the likelihood of a straightjacket ever being used to restrain a crewmember is remote; if a crew of mostly uneducated sailors could survive the deprivations of living in the Arctic for three years without coming to blows, then I'm sure a group of the most highly trained people on the planet will be able to survive surrounded by all the luxuries the 21st century has to offer. Then again, after the failures of ESA's tin can experiments, maybe not!

So, we've discussed several possible psychological problems that might be encountered by interplanetary astronauts, but so far, no show-stoppers. From a mission planner's perspective, it would seem the psychologists got it wrong, but what about depression? Now, you may be wondering why astronauts would get depressed. After all, surely these lucky crewmembers will be *positively* affected by their journey and the thrill of overcoming the challenges of traveling to and living on another planet. While it's true that crewmembers will experience these and other positive reactions, the issue of depression is common not only to polar exploration and long-duration space travel, but also to several other environments in which humans live for long periods of time. Fortunately, scientists at the National Space Biomedical Research Institute (NSBRI) are developing an interactive, multi-media program that will help astronauts recognize and effectively manage depression and similar psychosocial problems. The self-guided treatment is known as the Virtual Space Station (VSS) and was developed for NASA by Drs James Cartreine and Jay Buckey. The depression self-treatment has already been tried by researchers in Antarctica and, according to Cartreine, feedback has been positive. The VSS system is a self-guided computer program that uses an established behavioral-treatment approach known as problem-solving therapy. Here's how it would work. The depressed astronaut sits in front of a computer and is guided through the program's series of steps by a pre-recorded voice and an image of a psychologist. Then, the crewmember is instructed to make a list of measurable problems they're experiencing before choosing one problem from the list that has a reasonable likelihood of being solved and to create a goal that solves the problem. With the help of the computer, the crewmember chooses an action that will solve the problem. It's a step-by-step process in which the computer doesn't solve the problem, but facilitates the crewmember's own problem-solving behavior. The astronaut doesn't necessarily have to be depressed to use the VSS. Imagine you're on an eight-month-long flight, the crew isn't getting along, you're not getting enough sleep, and you're stressed out with nobody to talk to. What do you do? You log onto the VSS of course! Going to a computer for help with personal problems might seem strange, but research has shown that people are more willing to open up to computers than therapists. It also happens to be a common theme in science fiction. Those of you who are fans of the cult sci-fi series *Red Dwarf* will remember Kryten, a Series 4000 mechanoid, and Holly, the ship's 10th-generation AI hologrammatic computer, both of whom were considered part of the crew. In fact, Holly's user interface appears on ship screens as a disembodied human head on a black background – a design that isn't too far removed from the VSS. The computer–human interaction is a theme that is also often explored by Hollywood. Who can forget *2001: A Space Odyssey* and the relationship between David Bowman and the HAL 9000 and, more recently, the interaction between Sam and Gerty in Duncan Jones's classic film *Moon*:

> INT. REC ROOM -- MORNING 15
>
> A state of the art robot, a GERTY 3000 -- known simply as 'GERTY' -- is preparing Sam's breakfast. Gerty is in three sections and moves along a horizontal rail that runs throughout the base. He has a readout screen that perpetually spews data. His hands resemble pincers, but are perfectly nimble.
>
> For the purposes of helping run the base and looking after Sam 1, Gerty is as good as human, if not better.
>
> Sam 1 enters.
>
>   SAM 1
>
> Two weeks!
>
>   GERTY
>
> Morning, Sam. How are you today?
>
> Sam 1 grins.
>
>   SAM 1
>
> Hey, you want to give me a haircut later?
>
>   GERTY
>
> Of course. How's your headache?
>
>   SAM 1
>
> Much better, thanks, pal. Yeah. Good! A hair cut ... I think I'm going to shave too. What do you think, Gerty? Think it's time to get rid of this?

Sam 1 feels at his chin. He treats Gerty more like a person than a robot. Whether this is down to Gerty's intelligence or Sam 1's desperation for company isn't clear just yet.

Extract from the film *Moon*, by Nathan Parker, *www.script-o-rama.com*

MISSION OPERATIONS

Anyone with a basic knowledge of the history of polar exploration and a reasonable understanding of extreme and isolated environments will acknowledge that of all the problems faced by interplanetary astronauts, the spectrum of psychological problems will probably have the least impact. Nevertheless, despite the relatively benign impact psychological disorders may have upon an exploration class mission (ECM), there is always the Lisa Nowak or Skylab-4 variable. To guard against the unpredictable, space agencies will need to implement countermeasures.

The first countermeasure will be screening astronauts. Formal psychiatric examinations and psychological testing occur when candidates are screened in their application to become astronauts. ECM astronauts will be those who show above-average performance in the following areas: logical reasoning, mental arithmetic, memory function, attentional abilities, auditory and visual perception, spatial comprehension, psychomotor functions, psychomotor coordination, multiple task abilities, problem-solving abilities, decision-making abilities, and communication abilities. On the criterion of personality factors, weight will be given to motivation, dominance, empathy, aggression, stress-resistance, readiness to bear privations, work orientation, loyalty, and ethical integrity. Oh, and a sense of humor! It's a rigorous system that has worked well for selecting crews for increments onboard the ISS and Shuttle flights, but for interplanetary missions, potential crewmembers will probably be under special scrutiny for the aforementioned factors.

Another important countermeasure will feature pre-mission training between crewmembers and Mission Control personnel. No doubt, astronauts will be involved in some form of social competence training that focuses on the need for a collaborative effort during the mission. Special attention will also be given to interpersonal behavior and attitudes, training astronauts to develop empathy, social sensibility, and tolerance. Stress management will also feature in this pre-mission training, with special attention directed at identifying warning signs and the use of stress-coping techniques. Mission trainers will also spend time developing team-building skills, focusing on the key stages of *forming*, *storming*, *norming*, and *performing*. During the forming stage, the crewmembers try to find their position in the team and interpersonal contacts increase, while the storming phase serves to clarify individual roles, which is usually characterized by conflicts between crewmembers. During the norming stage, common goals and skills are defined – a process that ultimately allows the crew to perform the task. Assuming the mission planners have learnt from historical space psychology, special emphasis will be given to the role of the leader – as we've seen in polar exploration, the reason many men survived these expeditions was thanks to effective leadership. Given the high stress of ECMs, it is likely that the role of the leader will be elevated to an even more critical position. In fact, how an ECM commander relates to the mission goals and to the crew may literally make or break the expedition. I would suggest Charles Lansing's account (*Endurance*) of Shackleton's expedition be mandatory reading for ECM commanders! Last, but not least, training will include team-building experiences in spacecraft simulators using Crew Resource Management (CRM) and Line-Oriented Flight Training (LOFT) to develop crew coordination skills. CRM and LOFT may be augmented by problem-oriented team supervision (POTS) – a psychological intervention method in which the crew works together on actual crew-related matters that have emerged during their training routine (e.g. interpersonal problems).

Once en route, Mission Control will monitor the crew using crew–ground audio communications, observed video behavior, and analysis of speech patterns. The crew will also monitor themselves, using tools similar to the VSS. They will also avail themselves of private conferences with family and friends; some of these private conferences might be similar to the one portrayed in the movie *Moon*:

```
INT. COMMS NOOK
Sam 1 sitting before the monitor. He hits the 'PLAY' button,
begins watching the message.
ON THE MONITOR: Tess is sitting in a spacious living room
talking to Sam 1. Tess has a sweet voice, she sounds grounded,
like she's got a head on her shoulders.
                         TESS
        Hi Sam. It's me. How are you?
                      (a beat)
        I got your last message, it was really great to hear
        your voice. I miss you too.
                       (MORE)
        I know you've been really lonely up there, but in a
        lot of ways I think it's been good for you. For both of
        us. I hope you don't mind me saying that. I'm proud of
        you.
                      (a beat)
        Hey, someone's got something to say.
A WOMAN, possibly a nanny or some form of hired help, swings a
LITTLE GIRL into Tess's arms. Tess looks a little embarrassed
by this. Having a nanny is a new part of her life. The little
girl is EVE, Sam and Tess's daughter.
                     TESS (cont'd)
        What did you want to say to daddy, baby?
Eve just stares. Tess whispers to her ('Remember what we
```

Practiced', etc.) Finally Eve attempts:

                    EVE

  Asstraut.

                    TESS

  Who's an astronaut?
               (encouraging)
  Go on!

                    EVE

  Daddy asstraut!

Tess laughs. So does Sam 1.

                    TESS

  That's right, daddy's an astronaut. Clever girl!

Eve fidgets, rubs her nose, distracted.

                TESS (cont'd)

  She's shy. Uh, Cathy, could you . . .?

The nanny steps in, hoists Eve away. Tess waits until they're out of earshot.

                TESS (cont'd)

  I'm still not used to that -- and this house! It's
  amazing. Thank you.
                    (beat)
  I can't believe you are going to be back soon. It's
  her birthday next month. I thought we could pick out a
  play house for the garden.

                TESS (cont'd)
                    (MORE)
             (getting excited)
  We can pick it out together!

A pause. Tess just stares into the camera. She is hundreds and thousands of miles away, but for a second it feels like she's right there in the Comms Room with Sam. It's intimate.

She finally shakes her head, self-conscious, shy.

                TESS (cont'd)

  God, I hate these things. Sam, I love you. I'm
  thinking of you always. I can't wait to see you,
  sweetheart. Okay. Bye.

And the message ends.

    ON SAM: smiling, on the brink of tears.

Extract from the film *Moon*, by Nathan Parker, *www.script-o-rama.com*

No doubt about it, like Sam, astronauts will get homesick. But at least they will be able to talk to their family and friends, which will be a huge psychological boost. Remember, Shackleton's men didn't see their families for more than three years. Interplanetary astronauts will also have the opportunity to avail themselves of remote psychological counseling using two-way audio-visual links if required. The counseling sessions will be supportive sessions conducted by a specialist experienced in individual psychopathologies and small group behavior.

POST-MISSION MENTAL HEALTH CARE

Although much has been written on the subject of how astronauts might behave during the flights, what happens when they return from their three or four-year mission? While a NASA-sponsored longitudinal follow-up study of astronauts' health has not revealed any untoward psychiatric problems of participation in long (six-month)-duration missions, the stress of reintegration and post-flight adjustment has been noted. The unpredictable effects of mission-related physiological changes and prolonged exposure to radiation, coupled with the emotional stress of reintegration following an absence of three, four, or five years means a post-mission program of psychiatric assessment and family support will be imperative.

From the discussion of stressors and psychological issues, it is safe to say that interplanetary missions will expose crewmembers to many of the same risks as have been reported during polar exploration, Antarctic research stations, and onboard the ISS. Sure, there will be some new psychological problems that will arise during multi-year ventures such as exposure to radiation and the dependence of the crew on life-support systems without the possibility of rescue, but these don't warrant the clarion call for more preparatory psychological research. Interplanetary missions, in common with the expeditions of Nansen, Shackleton, and Amundsen, will be inherently risky. Crewmembers will become depressed, bored, irritable, and maybe even violent, but there is no behavioral or psychological threshold short of which today's current crop of astronauts is not ready to embark on such a mission. In fact, astronauts will always be ready if the risks are judged worth taking. The risks and unknowns faced by Nansen, Shackleton, and Amundsen and their crews were of similar magnitude to those faced by future long-duration astronauts, yet these great polar explorers were still willing to spend years away on expeditions characterized by horrific cold, limited supplies, and tuberculosis. No shirt-sleeve environment or pressurized habitat for them. Why, knowing the risks, did these men embark on such expeditions? Some went along for the money and, in part, for the fame, but mostly because the urge to explore was embedded so deep in them that the dark specter of death was no deterrent. The point is, despite the psychological and behavioral issues discussed here, humans are more than capable of surviving and staying sane on multi-year missions in horrendous conditions and in the face of appalling hardship.

Section III

Future Developments

7

Bioethics, sex, and cloning

Imagine a crew embarked upon a commercially funded interplanetary mission sometime in the (hopefully) not too distant future. Four months after launch, Dr Steele, the crew medical officer (CMO), performs his weekly examination of each crewmember and discovers that Danny Preston, the flight engineer, has a terminal illness. The crew is still three months from arriving at their destination, with no abort capability. What should they do? Should Preston be allowed to continue as a functioning crewmember, using up valuable life-support consumables, all the while knowing he is going to die? Should Preston sacrifice his life for the greater good of the mission? Does the commander even give Preston a choice? And what happens when Preston dies? Does the crew store his body until they return to Earth, do they bury it on the planet, or do they simply flush it through the airlock? Questions like these, and many more, must be answered before a crew departs on an exploration class mission (ECM). While science fiction has explored some of these questions, it probably isn't a good idea to rely on science fiction as a template for bioethical policy. But, before we start discussing what bioethical policy space agencies might adopt to deal with "who gets thrown from the lifeboat"-type scenarios, it's worth understanding what bioethics is.

Bioethics is a subject encountered practically every day in the news media, whether it's a discussion about cloning, an argument about stem cell research, or a debate about doping in sport. However, not all bioethical issues are categorized so easily. That's because there are different ethical systems, ranging from divine command theory, which dictates moral standards are set by God, to utilitarianism, which dictates the greatest good for the greatest number of people. In short, bioethics is a discipline in which there are no right or wrong answers and one that requires the use of critical thinking skills and the application of paradigms. Perhaps the best way to understand some of the difficult bioethical issues that may be faced by ECM crews is to analyze some hypothetical scenarios.

THE SURVIVOR SCENARIO

We'll start by imagining a four-man mission that's coming to the end of a three-

116 Bioethics, sex, and cloning

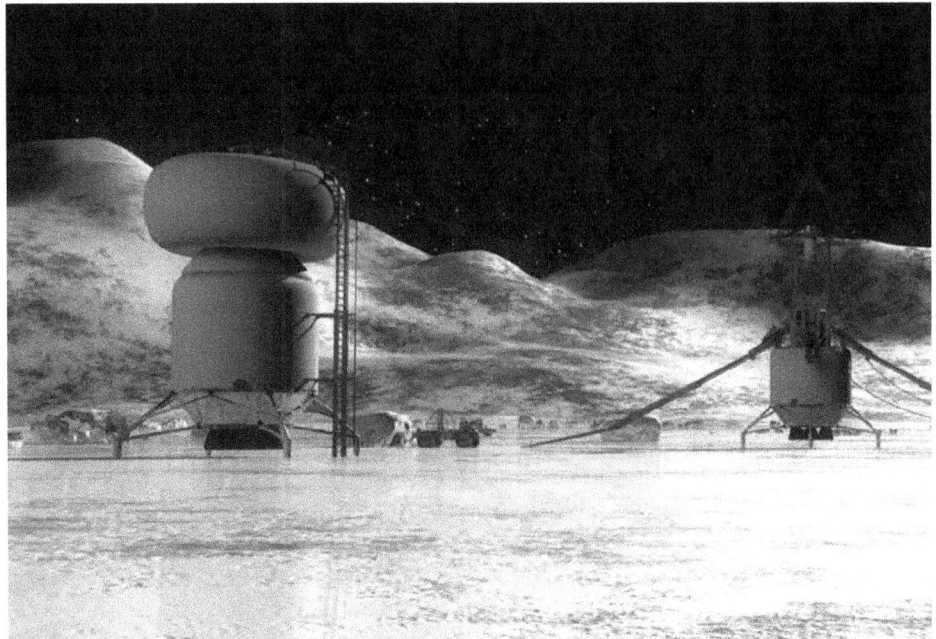

Figure 7.1 Artist's rendition of Callisto. In 2003, a NASA-led study identified revolutionary concepts and supporting technologies for a Human Outer Planet Exploration (HOPE) mission to Callisto, the fourth of Jupiter's Galilean moons. Assumptions for the Callisto mission included a launch year of 2045 or later, a spacecraft capable of transporting humans to and from Callisto in less than five years, and a requirement to support three humans on the surface for a minimum of 30 days. Image courtesy: NASA.

month stay on the surface of Callisto (Figure 7.1). The crew is busy preparing the surface lander and habitat for extended hibernation mode when they receive an emergency message from the orbiting crew vehicle that is due to take them home. A micrometeorite has punched a hole in the orbiting crew return vehicle. Before the automated self-repair nanobots can repair the hole, valuable life-support consumables have hemorrhaged away into space, leaving only sufficient oxygen for one crewmember to return to Earth. There is no rescue capability.

The four-man crew includes Commander Rains, 54, a stocky pilot and veteran of two Mars missions. Divorced with two kids, Rains is the son of a US Congressman. Luce, 49, the mission's flight engineer, is married with one child. Three weeks into the surface mission, it was her quick thinking that saved the crew from certain death when the air revitalization system broke down. Van Vogt, 36, the skinny co-pilot and the mission's youngest crewmember, is engaged to be married. A week ago, he learned that his mother has cancer. Reinauer, 62, the science officer, is the inventor of the propulsion system that made the Callisto trip possible in the first place. The details of the crew's predicament are passed on to the bioethics panel that is convened via an emergency conference call to choose who gets left behind on the surface.

Figure 7.2 The entry, descent, and landing (EDL) sequence includes phases such as cruise stage separation, parachute deployment, heat-shield separation, back-shell separation, and retropropulsion firing. Each event must occur within a very narrow operational envelope and most must be triggered autonomously, based on estimates of where the spacecraft is relative to the ground and how fast it is traveling. Furthermore, each event must be executed flawlessly in the presence of potentially significant variability in winds, atmospheric properties, and surface topography. Image courtesy: NASA.

THE TERRORIST ATTACK SCENARIO

After a four-month voyage, the crew of the first mission to Mars is three days away from performing the critical entry, descent, and landing (EDL) tasks when an Iranian-sponsored terrorist attack devastates Houston, the home of four of the six crewmembers. The families of two of the crewmembers are immediately confirmed killed in the attack, the wife of the pilot has been severely injured, and the children of the mission's flight engineer have been killed. What does Mission Control do? Do they tell the crew about the attack and not mention the dead relatives? Do they tell them about the dead relatives and hope they can still perform the EDL (Figure 7.2) tasks? Do they choose not to tell them anything at all until they are safely on the surface of Mars? Or, do they choose to keep the attack secret until the crew returns to Earth in six months?

THE INJURED CREWMEMBER SCENARIO

It's the year 2036 and the United States becomes the third nation to land on Mars (after China and Russia). Leading the four-man crew is Commander Krauthammer, the primary pilot and an avid mountaineer. The flight surgeon is Dr Riga, who is also the life sciences expert and second-in-command. Rounding out the crew is Carter, the flight engineer, and Floyd, the geophysicist. They've been exploring the surface for more than four months, collecting data and rock samples from the local area. Fifteen kilometers from the landing site is a 3,000-m peak that reminds Krauthammer of a mountain he climbed in Absaroka National Park. Krauthammer's been eyeballing the mountain for weeks, itching to have the opportunity to scale a virgin peak. On one of the mission's few scheduled rest days, he asks if anyone is interested in climbing the Martian mountain. Riga's feeling tired, so she declines, but Floyd and Carter are game. It takes them six hours to scale the mountain. Everything seems to be going well until Carter loses his footing on a scree slope. He tumbles several hundred meters before a boulder stops his fall. Fortunately, his robust Bio-Suit is intact, but the impact renders Carter unconscious. By the time Krauthammer and Floyd reach the flight engineer, Carter is hemorrhaging blood from a gaping head wound. The commander and geophysicist manage to carry him back to the base, where Riga examines Carter, a father of three. Riga only has limited medical facilities, but she's able to make a diagnosis, which is grim. Carter has a fractured skull and has suffered possible brain damage. If he had been on Earth, he would have been admitted to an intensive care unit. Riga gives Carter a less than 25% chance of survival. Krauthammer, wishing he had never suggested climbing the mountain, sends the accident report to Mission Control requesting guidance. The accident report takes half an hour to reach the flight surgeons and another two hours go by before they send their advice to Krauthammer.

Carter's prospects are grim. Orbital mechanics dictate the earliest return is several weeks away. Even with the Variable Specific Impulse Magnetoplasma Rocket (VASIMR – see Chapter 2) propulsion operating at full burst, the journey home will take five weeks and Riga expects Carter to survive no more than a week. Although Carter is sedated, he requires full-time medical attention, which includes having his vital signs monitored, being fed, and being moved. The crew has full-time jobs, which means providing round-the-clock care for Carter is next to impossible. Also, as the flight surgeons point out in their assessment of Carter's brain scans, while Carter may regain consciousness, it is likely he has sustained serious brain damage, and will not be able to function as an able-bodied member of the crew. Krauthammer retreats to his cabin and is about to deliberate the situation when he receives more bad news: Carter's wife has found out about the accident and has been interviewed by Fox News. Fox is running a story about how the space agency is going to leave Carter to die alone on the surface of Mars. Meanwhile, Reinhold Saberhagen, chairman of the House Committee on Science, has been informed of the events taking place on the surface of the Red Planet. He's understandably concerned, because it was his committee that approved the $30 billion mission, and he's especially uneasy about the possibility the agency might euthanize Carter; a death onboard an American spacecraft would not only give the Chinese (and Fox News!) a field day, but also turn the American public away from funding future missions.

THE SOLUTIONS

The solutions? Remember, there are no right or wrong answers in the field of bioethics. Only recommendations. First, let's consider the survivor scenario. This is probably the easiest to resolve. Since the crew return vehicle can only be operated by a pilot, the choice is between Rains and Van Vogt. How would you choose? Well, the critical factor is life-support consumables, so it would make sense to choose the crewmember with the lowest metabolic consumption of oxygen. After all, it's a long way from Callisto to Earth (628,743,036 km on a good day) and you want to make sure that whoever you choose has the best chance of making it back alive. Based on this rationale, it would seem that Van Vogt would be saying a painful goodbye to Rains, Reinauer, and Luce. But, don't feel too bad for them; they would be able to generate oxygen thanks to an in-situ resource utilization (ISRU) system and hopefully, after a few years, another spacecraft might visit Callisto and rescue them.

The terrorist attack is a little more complicated. Prior to the mission, crewmembers would have signed contracts binding them to an agreement that if a catastrophic event were to occur, the mission would take priority. More than likely, Mission Control would withhold the details of the attack until the crew had safely landed on Mars. Then, after the crew had performed post-landing operations, Mission Control would send an encrypted message to the commander and leave it up to him to decide how much information to tell the crew.

Now, what about the injured crewmember? Well, despite the protestations of Carter's wife, the sensationalism of Fox News and the concerns of Mr Saberhagen, the mission would continue and Carter would probably be allowed to die. From a bioethics perspective, such an act would fly in the face of the principle of *nonmaleficence*, which obliges doctors to refrain from causing harm to their patients. But remember, Carter's prospects are grim and he's not expected to live much longer. In such circumstances, Dr Riga would have no obligation to provide pointless and futile treatment. You see, this is a situation in which the patient has reached a point at which further treatment is hopeless and therefore becomes optional. If he had been on Earth, Carter would probably receive palliative treatment, but the nearest intensive care unit is more than 54 million kilometers away. But simply withholding treatment doesn't mean Carter is going to die immediately. He could hang on for days. So, what if the crew agrees that Carter is using up valuable life-support consumables and decides to terminally sedate Carter? Wouldn't this be murder? No, it wouldn't, because an ECM would require crewmembers to agree to refuse treatment (or even be euthanized) to bring about their deaths under grim circumstances such as Carter's. Such justified assisted suicide would be outlined in a document similar to an advanced health care directive, also known as a living will or advance directive. An advance directive is given by individuals specifying what actions should be taken for their health in the event that they are no longer able to make decisions due to illness or incapacity, and appoints a person to make such decisions on their behalf. A living will is a type of advance directive that leaves instructions for treatment. One example of a document that combines the advance directive with a living will is the Five Wishes (Panel 7.1) advance directive.

> **Panel 7.1.** Five Wishes
>
> Five Wishes was a document introduced in 1998 that combined a living will and health care power of attorney in addition to addressing matters of comfort of care. The wishes are as follows:
>
> Wish 1: The Person I Want to Make Care Decisions for Me When I Can't. This section assigns a health care agent who makes medical decisions on your behalf if you are unable to speak for yourself. During an ECM, this would probably be the Commander.
>
> Wish 2: The Kind of Medical Treatment I Want or Don't Want. This section is a living will defining what life-support treatment means to you, and when you would and would not want it. For ECM astronauts, there wouldn't be much choice in the matter.
>
> Wish 3: How Comfortable I Want to Be. This section addresses matters of comfort care, such as what type of pain management you would like.
>
> Wish 4: How I Want People to Treat Me. This section speaks to personal matters, such as whether you would like someone to pray at your bedside, among others.
>
> Wish 5: What I Want My Loved Ones to Know. This section deals with how you wish to be remembered and final wishes regarding funeral plans. Again, for ECM astronauts, the decision will have already been made; if you die en route, your body will be ejected through the airlock.

Before leaving on the mission, astronauts will probably submit a form similar to an advance directive or perhaps a version of the Five Wishes. Alternatively, they may complete a document that combines the topics addressed in a living will with those of a Health Care Power of Attorney (HCPOA), although there would be some differences between the space agency HCPOA and the one the average citizen signs. For example, a standard HCPOA allows you to name someone (an Agent) to make health care decisions for you if you are unable to do so. Like a living will, a HCPOA usually allows you to state your wishes about certain medical procedures. Of course, an ECM astronaut's agent would most likely be the commander or doctor and they probably would have to rely on similar documents before deciding on which medical options to implement. Such a document would be an Advance Health Care Directive, since it combines the features of a living will and a HCPOA. The HCPOA would only go into effect if an astronaut was injured so badly that they were rendered permanently unconscious and unable to communicate.

Most probably, Carter (if he regained consciousness) would volunteer to be euthanized and his body buried on the surface of Mars (the first Martian burial), although his family would probably request that his body be returned to Earth – a request that would be denied for obvious reasons.

Having reviewed some possible ECM scenarios, it's obvious that current ethical standards are woefully inadequate for crewmembers embarked upon multi-year missions. That isn't really surprising though, because current ethical standards for astronauts were developed in an era of short-duration space missions when repeat missions were the norm and a return to Earth within days or hours was possible. In future missions beyond Earth orbit, a diverse group of astronauts will venture to remote destinations for increasingly long periods. Contact with Mission Control will be delayed and rapid return impossible. Multi-year missions to the moons or the outer planets will inevitably create special circumstances for which current ethical standards are inadequate. As the prospect of multi-year missions approaches, mission planners will design a new ethical framework to guide mission commanders and crewmembers in their decision-making when it comes to dealing with some of the potentially awkward moral questions such as the ones discussed earlier. One step towards achieving this is NASA's appointment of Dr Paul Wolpe, the agency's first Chief Bioethicist. Wolpe has already had experience tackling thorny bioethical problems. For example, when the Space Shuttle *Columbia* disaster occurred, one of the first questions NASA officials asked the University of Pennsylvania professor was if the government of Israel would request genotyping of all the human remains to separate Ilan Ramon's (the first Israeli astronaut) tissues from the others – a practice in keeping with Jewish burial tradition. It took Wolpe a while to work through the logistical problems such a request would entail but, ultimately, the Israeli government never made the request and the end result was that every grave contained some remains of each crewmember to commemorate their collective sacrifice in the name of space exploration.

NO SEX PLEASE, WE'RE ASTRONAUTS

Perhaps the most discussed ethical quandary is the issue of how to cope with sexual desire among a crew of healthy young men and women. The stories about whether astronauts have had sex in space are legion. In February 2000, allegations were made that Russian cosmonauts had conducted sex experiments in low Earth orbit (LEO) and that NASA had plans to conduct similar tests. French science writer, Pierre Kohler, author of *The Final Mission*, alleged the experiments were described in a secret NASA publication identified as No. 12-571-3570, a document that was discovered – surprise, surprise – on the internet. NASA responded by saying no legitimate document corresponding to that number or subject matter existed and that NASA had never had pursued such research. Unfortunately for Kohler, who built his case based on facts supposedly contained in the mysterious NASA document, there's no evidence it ever existed. While the allegation has since been relegated to the status of internet hoax, readers might be interested in an edited version of the document that was circulated in 1995:

Experiment 8 Postflight Summary: NASA publication 12-571-3570

Introduction
The number of married couples currently involved in proposals for long-term

projects on the US space station has grown considerably in recent years. This raises the serious question of how such couples will be able to carry out normal marital relations without the aid of gravity.

Preliminary studies in the short-term weightless environment provided by aircraft flying on ballistic trajectories were sufficient to demonstrate that there were problems, but the duration of the zero-G environment on such flights is too short to reach any satisfactory conclusions. Similar experiments undertaken in a neutral buoyancy tank were equally inconclusive because of the awkwardness of the breathing equipment.

The primary conclusion that could be drawn from these early experiments was that the conventional approach to marital relationships (sometimes described as the missionary approach) is highly dependent on gravity to keep the partners together. This observation leads us to propose the set of tests known as STS-75 Experiment 8.

Methodology
The co-investigators had exclusive use of the lower deck of the shuttle for 10 intervals of 1 hour each during the orbital portion of the flight. A resting period of a minimum of 4 hours was included in the schedule between intervals. During each interval, the investigators erected a pneumatic sound deadening barrier between the lower deck and the flight deck (see NASA publication 12-571-3570) and carried out one run of the experiment.

Each experimental run was planned in advance to test one approach to the problem. We made extensive use of a number of published sources in our efforts to find satisfactory solutions, arriving at an initial list of 20 reasonable solutions. Of these, we used computer simulation (using the mechanical dynamics simulation package from the CADSI company) to determine the 10 most promising solutions.

Six solutions utilized mechanical restraints to simulate the effect of gravity, while the others utilized only the efforts of the experimenters to solve the problem. Mechanical and unassisted runs were alternated, and each experimental run was videotaped for later analysis. Immediately after each run, the experimenters separately recorded their observations, and then jointly reviewed the videotapes and recorded joint observations.

The sensitive nature of the videotapes and first-hand observations precludes a public release of the raw data. The investigators have prepared this paper to summarize their results, and they intend to release a training videotape for internal NASA use, constructed from selected segments of the videotapes and additional narrative material.

The report went on to describe mechanical solutions and natural approaches that utilized elastic belts, harnesses and various other devices, a discussion of which is beyond the scope of this book and best left open to the reader's imagination.

While NASA hasn't discussed the problem of sex in space (see Figure 7.3), it does acknowledge that policies need to be developed to address the issue. Of course, some argue that sex isn't any of NASA's business. Others suggest that there is very little

No sex please, we're astronauts 123

Figure 7.3 In September 1992, two married astronauts, Jan Davis (pictured) and Mark Lee, flew aboard STS-47. NASA's policy normally prohibits married couples flying together, not because they're afraid they'll have sex, but because it might hurt the team dynamic. However, the agency made an exception for Davis and Lee, since the couple got married so close to launch time. While Davis and Lee have refused to answer questions about the nature of their relationship (they later got divorced) during the mission, it's possible the United States lost the "sex in space" race because there have been rumors that Elena Kondakova and Valery Polyakov may have joined the 200-mile-high club. Image courtesy: NASA.

NASA can do once the crew leaves LEO anyway. But what about the potential negative consequences of pregnancies en route? Do we sterilize ECM crews? At polar research stations, there is sexual contact between men and women and those missions seem to function just fine. How do they do it? Well, first of all they use a lot of condoms – thousands of them; before the six months of winter darkness descends over Antarctica's McMurdo Station, the research base takes delivery of 16,500 condoms. There's also an unspoken behavior at the base where some crewmembers take a spouse for the time that they're there – they have an exclusive relationship with someone during the deployment and it's understood that when they leave, the relationship is over. Of course, whether such a system would work in space would depend on whether the ECM crew was all-male, all-female, or co-ed. Up until the 1980s, NASA crews were all-male. Then, in 1983, Sally Ride became the first American woman in space aboard the Shuttle *Challenger* and since then, missions have routinely included one or two women. Now, you may think that the problem of sex in space may resolve itself because crews will have heavy workloads and little play time. No doubt about it, astronauts are very task-focused, but during a three or five-year mission, it is inevitable that sexual frustration will ensue. That's because human sexuality is a basic need and it doesn't matter if the space agency tells their astronauts "Hey, for three years, you can't do that". They're going to figure out a way to do it, and if they do it now, they're on their own because there is no official policy. But, while space agencies have traditionally stuck their collective heads in the sand on the issue of sex, it's unrealistic to ignore the psycho-social impact of sexual activity on space operations and the potential physiological consequences. Then there are the questions regarding issues related to sexual activity in space, such as behavioral health, team dynamics, pregnancy, and embryogenesis. For example, what impact will an in-flight sexual relationship have on team dynamics and efficiency? What are the chances of a successful pregnancy and delivery? What about the risk of ectopic pregnancy, miscarriage, or other complications? The list goes on.

When sexual relationships among a space crew become acknowledged, the medical system will have to deal with these and other questions. Training for both ground and flight crews will need to be adapted to address these issues. Perhaps the most pressing issue will be the psycho-social implications of in-flight sex. While space crews will be relatively small in number for the foreseeable future, if coupling occurs, it can have serious ramifications on the crew's working relationships, and therefore on mission success. After all, we know from Lisa Nowak's case that even professional astronauts on active flight status can develop serious mental health issues related to interpersonal relationships and the prolonged stressors of an ECM will only make such situations worse. And what about the issues facing the partner left behind on Earth if an adulterous relationship develops within the crew? What will the implications be for the other crewmembers? What procedures will they follow? Can they discuss the affair during their daily private medical conference (PMC)? Will Mission Control ask them to conceal the affair? What happens if the slighted partner on Earth starts divorce proceedings? How would *that* affect the mission? What would Fox News say? Perhaps the best way for space agencies to address the issue of sexual desire is by simply selecting an all-male or all-female crew. While such a policy might deselect better qualified candidates, mission planners may decide that the behavioral issues sex poses on a multi-year mission outweigh the disadvantages. Now, some may argue that an all-male crew would create problems, but precedents have been set by several multi-year polar expeditions. For example, Shackleton's Imperial Trans-Antarctic Expedition provides more than enough evidence that spending time with a group of men in a confined environment for several years under extraordinary stress doesn't result in any mission-compromising problems. However, let's face it, sooner or later, like hunger and thirst, sex is a biological motive and it doesn't make sense that men and women embarked upon ECMs are going to have no thoughts of it for several years. It's inevitable that crewmembers will engage in sex, so the agencies might as well pull their heads out of the sand and discuss the problem.

While an interplanetary affair will obviously cause problems, what about the implications for procreation? It's known that spermatogonia are radiosensitive and it's also known that ovarian function can be adversely affected by environmental conditions and the space environment isn't the healthiest environment for a pregnant woman. For one thing, the closed loop cabin environment includes all sorts of toxins, there's the risk of rapid decompression, and little is known about the effects of microgravity on embryogenesis. From research conducted onboard the International Space Station (ISS), we know that astronauts and rats exposed to microgravity have experienced endocrine imbalances, such as hypothyroidism. Such abnormalities in a pregnant woman could have significant effects on the fetus. For example, the development of the brain is regulated by thyroid hormones, and deficiencies in the neonatal brain can lead to cretinism and other abnormalities. Imagine caring for a retarded baby on the surface of Mars.

Another risk the ethics committee has to consider is the effect of space radiation upon embryogenesis. All astronauts are classified as radiation workers because they are exposed to so much radiation and when crews start venturing on ECMs, they will be exposed to even more radiation. Current guidelines recommend pregnant women be exposed to very little radiation because it is known that even small exposures can lead to fetal defects, such as retardation. We know that radiation causes DNA damage, cell death, and chromosomal abnormalities. Even if the fetus can develop normally in the space environment, many potential dangers remain. There are the conditions of pre-eclampsia, gestational diabetes, and Rh incompatibility, for example. How would the CMO treat these on the surface of Mars or Callisto? Compounding the problem is the fact that, based on astronaut demographics, the pregnant crewmember will be older than average, which means the pregnancy would already be classified as high-risk. The bottom line is that no one would want to become pregnant on an ECM, but it's best to be prepared, so it's more than likely pregnancy test kits will continue to be flown as part of the standard medical kit. Additionally, an onboard ultrasound kit would be useful to diagnose and document a pregnancy, as will non-surgical treatments, such as methotrexate, for the termination of ectopic pregnancies, which brings us onto the subject of in-flight terminations. Should mifepristone (a synthetic steroid compound used as an abortifacient in the first two months of pregnancy) be in the medical kit? Given the real concerns about the negative effects of the space environment, what ethical considerations must be addressed before terminating the pregnancy?

For many, the first question if a member becomes pregnant is whether to abort the mission, but in practice, this would be close to impossible, since an ECM would have very limited abort capability. For example, a Mars mission might have a single means of evacuation and if this is used to evacuate a single crewmember, the rest of the team would have no way to abandon the surface if another emergency arose. Therefore, would it be appropriate to abandon the entire mission just to bring home a pregnant woman? Not likely.

One side of the argument would contend that even if there is a higher-than-average risk of fetal malformations, there is an ethical obligation to return the mother to Earth as quickly as possible, regardless of the cost to the mission or the risk to the crew from implementing an abort from the surface. The other, and more rational, side of the argument would maintain that such an abort would take months and probably overwhelm the onboard medical system (intravenous fluids and analgesics might need to be rationed or withheld in preparation for an obstetrical event, rather than being available for use during nominal mission activities) to the detriment of the remaining crew. And, what if the pregnant crewmember was one of the designated extra-vehicular crewmembers? That role would have to be transferred to her backup, thereby compromising that crewmember's duties. Furthermore, as the pregnancy progresses, there could be additional issues such as the landing. A heavily pregnant crewmember is unlikely to fit in her seat liner, let alone be able to withstand typical G forces associated with landing. And what if there were complications? The list goes on. Even assuming the pregnancy is green-flagged (extremely unlikely) to continue on the surface, how would childbirth take place? Should the medical system be designed with a Martian pregnancy in mind? And what about caring for the first Martian (technically an alien)? Would the crew have to bring along baby clothes and diapers? Which part of the landing vehicle would be turned into a day-care center? Let's face it, pregnancy during an ECM is not a good idea but, like sex in space, it's a problem space agencies will eventually face. Will a baby born in space be healthy? Well, a pregnant woman has yet to travel in space, but some studies have flown pregnant rats in space. In 1983, the Soviet Union launched a satellite with a pregnant rat on board and found the trip was harder on the mother than on the fetuses. Once the babies were born on return to Earth, they were slightly thinner and weaker than their Earth-based counterparts and lagged behind in their mental development, although they eventually caught up. More recently, in 2001, NASA sent pregnant rats into space to measure some of the effects of the microgravity environment on the fetuses. The rats were sent in the middle of their pregnancies when the vestibular systems (a network of channels and sacs of fluid in the inner ear that regulates balance) were beginning to develop in the fetuses. The mothers and babies fared better in this experiment than in the 1983 study. The mothers gave birth to normal-sized babies and were able to care for them normally, although there were noticeable effects on the vestibular systems of the space-based rat infants. In contrast, the

Earth-based babies were able to immediately right themselves upon being turned on their backs in water, whereas the space-based babies had more trouble, although after five days of the same test, all the babies were able to roll over. The researchers also found that the vestibular organs detecting angular changes were actually more advanced in the space-based babies, probably because their mothers were forced to roll around a lot on the Shuttle due to the lack of gravity. So, while we're still a long way from determining whether a human can give birth in space, the findings so far seem promising.

PULLING THE PLUG

Another pressing ethical question concerns what action to take if an astronaut becomes terminally ill during a mission. In such an event, the Commander may be directed by Mission Control to euthanize the ill crewmember in order to preserve medical supplies and life-support consumables. Alternatively, the affected crewmember, knowing he/she has only a short time to live, may offer to sacrifice his/her life for the mission (see Figure 7.4). In such a situation, what will mission guidelines instruct the Commander to do? Needless to say, euthanizing a crewmember will not endear a space agency to the media (Fox News will have a field day!) or the public, who assume an astronaut's well-being will take precedence over mission success. Such a perception is not surprising, since, to date, any astronaut becoming ill or injured onboard the ISS has had the opportunity to simply leave the outpost onboard the Soyuz and return to Earth within hours. Unfortunately, this will not be possible when the nearest hospital is several million kilometers away! For situations such as "who gets thrown from the lifeboat", it will be necessary to equip mission commanders and crew with the necessary ethical framework (see Appendix) to make difficult decisions.

THE ONE-WAY TRIP OPTION

> "Did the Pilgrims on the Mayflower sit around Plymouth Rock waiting for a return trip? They came here to settle. And that's what we should be doing on Mars. When you go to Mars, you need to have made the decision that you're there permanently. The more people we have there, the more it can become a sustaining environment. Except for very rare exceptions, the people who go to Mars shouldn't be coming back. Once you get on the surface, you're there."
>
> Buzz Aldrin, *Vanity Fair*, June 25th, 2010

In the same way as European pioneers headed to America knowing they would not return home, some argue that the first astronauts sent to Mars should be prepared to spend the rest of their lives there. That's what Buzz Aldrin proposes, and he isn't the only one. If you think such a mission is a little extreme, consider the race to the

128 Bioethics, sex, and cloning

Figure 7.4 If you think self-sacrifice is an extreme option, consider the case of Captain Oates. Oates was an English Antarctic explorer, known for the manner of his death. In 1910, Oates applied to join Robert Falcon Scott's expedition to the South Pole. Scott eventually selected him as one of the five-man party who traveled to the Pole. On January 18th, 1912, 79 days after starting their journey, Scott's party reached the Pole, only to discover Norwegian explorer Roald Amundsen had beaten them to it. Scott's party faced extremely difficult conditions on the return journey, and the effects of scurvy and frostbite slowed their progress. Oates' feet became severely frostbitten and he weakened faster than the others. His slower progress, coupled with the unwillingness of his three remaining companions to leave him, caused the party to fall behind schedule. On March 15th, Oates told his companions he couldn't go on and proposed they leave him in his sleeping bag, which they refused to do. He managed a few more miles that day but his condition worsened. Waking the next morning and recognizing the need to sacrifice himself to give the others a chance of survival, Oates told his companions "I am just going outside and may be some time". Then, he walked into a blizzard to his death. His death was seen as an act of self-sacrifice when, aware that his ill-health was compromising his three companions' chances of survival, he chose certain death. Image from the Photographic Archive, Alexander Turnbull Library. Image courtesy: National Library of the Government of New Zealand.

Moon in the late 1960s. Given the mundane state of manned spaceflight as it exists now, it's hard to imagine that in 1967, the Americans and the Russian were literally dying to get to the Moon. That year, three American astronauts lost their lives in the Apollo 1 fire, and a Soviet cosmonaut, Vladimir Komarov, died when the re-entry parachute of his Soyuz craft failed. Several more died during training. At the beginning of 1968, no one knew for sure which of the two nations would reach the Moon first, or whether the men they sent would return. Some wondered whether that was even a requirement. In the mid 1960s, NASA developed the vehicle designed to carry American astronauts to the Moon sooner rather than later. Such was the urgency that there were some who suggested the astronauts might risk a one-way trip to ensure that the first dead body on the Moon would be American. Not surprisingly, this wasn't the sort of thing NASA had in mind, but it didn't stop science fiction writers[1] or Hollywood, who thought the idea would be great material for a movie. In Robert Altman's 1968 film, *Countdown* (based on a novel by Hank Searls), astronauts James Caan and Robert Duvall compete for the honor of steering a modified Gemini spacecraft on a one-way trip to the Moon. Actually, this part was accurate because NASA had studied lunar missions involving the Gemini. Fitted with an extra translunar stage, a modified Gemini spacecraft could have sent two astronauts around the Moon by 1965, without Apollo hardware and at a fraction of the cost. Open-cockpit landers could have transported astronauts to the lunar surface by 1966. In *Countdown*, the plan was for the astronaut selected for the mission to camp out on the Moon until a later crew collected him. However, when NASA hears that the Soviets have launched first, Caan's character still wants to go and when he can't locate the shelter, decides to land anyway. In real life though, a suicide mission wasn't necessary, because by December 1968, the Apollo 8 astronauts had orbited the Moon, proving that humans were not only ready to go to the lunar surface, but also had a pretty good chance of coming back.

Before you discount the idea of an interplanetary one-way trip, remember that most of the danger of spaceflight lies in launching and landing, as the *Challenger* and *Columbia* disasters demonstrated so horrifically. By eliminating the return trip, you would cut the risk of the mission. And let's not forget Shackleton, Mawson, Amundsen, and company, who set out on their expeditions knowing full well they could die in the process. Why should it be different today? Now, of course, NASA would never entertain the idea of a one-way trip and it's unlikely that Congress would take the political risk and be willing to sign what might effectively be a death warrant for American citizens. But what about commercial enterprises and more adventurous space agencies? It's an intriguing idea that definitely merits attention.

First of all, there's the money saved from not having to ferry astronauts back to

---

[1] In Stephen Baxter's novel, *Titan*, a survivor of a space shuttle crash, Paula Benacerraf, a visionary JPL scientist, Rosenberg, and a former Moonwalker who wants to relive his glory years, Marcus White, come up with a plan to launch a manned, one-way mission to Titan using the remaining shuttle fleet and vintage Apollo spacecraft and Saturn V launchers.

Earth. Don't forget, whenever you want to bring astronauts back, you have to bring along the fuel for the return journey, which means you have to expend extra fuel just to haul that extra fuel. Then, you have to design a means of launching from the planet and provide life-support consumables to the astronauts for the journey back. And, of course, all the way back to Earth, the crewmembers are taking another big radiation hit. Chances are, they may not even survive. Talk to mission planners and engineers at NASA and they'll tell you that it would cost about 10 times more for a round-trip mission than a one-way trip.

> "Men wanted for hazardous journey. Low wages, bitter cold, long hours of complete darkness. Constant danger. Safe return doubtful. Honour and recognition in event of success."
>
> Advertisement rumored to have been posted (possibly in the December 29th, 1913, issue of *The Times*) by Sir Ernest Shackleton before his 1914 Imperial Trans-Antarctic Expedition

Now, you may be thinking how you would recruit crewmembers for a one-way ticket, but there's actually a long tradition of similar trips in human history. Colonists and pilgrims never expected to return. And the ad (it may be apocryphal, but its content applies equally to those selected for an interplanetary mission) that Ernest Shackleton placed in search of a crew for his 1914 expedition to Antarctica resulted in 5,000 applicants. That's more than applied for NASA's Class of 2009 astronauts! Let's not forget, the one-way explorers would be equipped with plenty of materials. They'd have a small nuclear reactor and a couple of rovers, they could make their own oxygen, grow food, and even initiate building projects using local raw materials. Supplemented by shipments from home, the colony could be sustained indefinitely, and eventually become self-sufficient. Sure, the living conditions would be cramped and uncomfortable, but so was the Antarctic for explorers a century ago.

Okay, so there might not be any problem recruiting people, but there is still the political football of sending people on a one-way trip. Part of the problem is overcoming the persistent mentality that a mission is only over when the astronauts land safely on Earth. For most people, anything else is failure, but it shouldn't be. Rather, a one-way mission should simply be part of a long-term plan of exploring the solar system. And let's not forget, these pioneers would be establishing the first human off-world colony – that's hardly defeatist, is it? Anyway, if you recall the hazards of radiation described in Chapter 4, you'll remember that an interplanetary trip may well reduce the astronaut's lifespan, probably as a result of increased cancer risk from being exposed to so much radiation. So, why waste the remainder of their lives dying on Earth in a hospital when they could be enjoying themselves and performing useful science on the planet?

But what about the risk? Surely, from an ethical perspective, it would be irresponsible to send a crew on such a solo mission. Well, people aren't averse to taking risk. Just ask someone who's climbed K2 (Figure 7.5), a mountain with a one in three chance of survival. Let's face it; the human species is genetically programmed to take on risky challenges.

Figure 7.5 K2. Image courtesy: Wikimedia Commons.

GATTACA

Another issue that will involve space bioethicists will be genetic screening – a subject discussed briefly in Chapter 3. While genetic screening[2] for disorders for which a successful therapy exists has been in place for many years, the practice still raises many questions. One of the problems posed by recent developments in molecular genetics is the definition of what constitutes a "disease".[2] For example, what happens if an astronaut candidate was diagnosed with an increased susceptibility to cancer, but has lived cancer-free all his/her life? Does this mean he/she has a disease?

Another problem is the question of what happens to all that information. For example, what happens to an astronaut candidate who is deemed not to have the right stuff in their DNA for an ECM? How can their genetic profile be kept confidential and how can the discriminatory use of the test results be prevented? Also, will the astronaut candidate have the right to choose not to know about their genes? And, who counsels the candidate whose genetic screen reveals they have an inherited disease which means they only have five years to live? Should laws be passed to protect people against genetic discrimination by private entities? What does Hollywood have to say? Here's a scene from the classic science fiction movie, *Gattaca*:

---

[2] Less than half of all disease and disability is thought to be caused by genetic factors and each human is thought to carry about five recessive genes for lethal disorders.

EXT. GENETIC COUNSELLING OFFICE BUILDING. DAY.

ANTONIO, MARIA and 2-YEAR-OLD VINCENT exit a packed commuter bus and enter a Genetic Counseling office building bearing the sign — 'PRO-CREATION'.

INT. GENETIC COUNSELLING OFFICE. DAY.

A GENETICIST stares into a high-powered microscope as ANTONIO, MARIA and 2-YEAR-OLD VINCENT are shown into the office by a NURSE. On the counter beside the Geneticist is a glass-doored industrial refrigerator containing petri dishes arranged on racks several feet high.

GENETICIST
(to the nurse, without taking his eyes from his binocular microscope)

Put up the dish.

While Antonio and Maria take a seat in front of a television monitor, the Nurse puts a labeled petri dish under a video-equipped microscope. The Geneticist swings around in his chair to greet his clients.

Four magnified clusters of cells — eight cells on each cluster — appear on the television screen.

GENETICIST

Your extracted eggs ... (noting the couple's names from data along the edge of the screen) ... Maria, have been fertilized with ... Antonio's sperm and we have performed an analysis of the resulting pre-embryos. After screening we're left with two healthy boys and two healthy girls. Naturally, no critical pre-dispositions to any of the major inheritable diseases. All that remains is to select the most compatible candidate.

Maria and Antonio exchange a nervous smile.

GENETICIST

First, we may as well decide on gender. Have you given it any thought?

MARIA
(referring to the toddler on her knee)

We would like Vincent to have a brother ... you know, to play with.

The Geneticist nods. He scans the data around the edge of the screen.

                        GENETICIST

You've already specified blue eyes, dark hair and
fair skin. I have taken the liberty of eradicating
any potentially prejudicial conditions — premature
baldness, myopia, alcoholism and addictive
susceptibility, propensity for violence and
obesity —

                         MARIA
                 (interrupting, anxious)

— We didn't want — diseases, yes.

                        ANTONIO
                   (more diplomatic)

We were wondering if we should leave some things to
chance.

                       GENETICIST
                      (reassuring)

You want to give your child the best possible start.
Believe me, we have enough imperfection built-in
already. Your child doesn't need any additional
burdens. And keep in mind, this child is still you,
simply the best of you. You could conceive naturally
a thousand times and never get such a result.

> Extract from the screenplay *Gattaca*, by Andrew M. Niccol,
> *www.script-o-rama.com*

Gattaca, released in 1997, is set in the not too distant future. In the film, Gattaca is a NASAeasque organization colonizing planets in the Earth's solar system. The organization has strict hiring policies, and accepts only genetically engineered human beings called "Valids". Vincent (played by Ethan Hawke) dreams of being an astronaut, but can't pass Gattaca's strict hiring policies because he isn't genetically engineered. As a person born via normal biological methods, Vincent is an "Invalid" – a designation alluding to the fact that in this future, genetic engineering has become the norm, and individuals born "normally" are considered inferior. To beat the system, Vincent borrows the identity from Jerome, a genetically engineered Valid, paralyzed from a car accident. Vincent borrows Jerome's genetic identity (blood, urine, and hair samples) and illegally enters the exclusive world of Gattaca. With Jerome's genetic signature, Vincent is able to join a prestigious mission to explore Titan, a moon of Saturn. In essence, *Gattaca* is a morality play, and its central theme is the morality of genetic engineering. In the discussion between Vincent's parents and the geneticist, the geneticist suggests that for an extra $5,000, he could give the embryo enhanced musical or mathematical skills – a procedure that would splice in a gene that was not present in the parents' original DNA. Perhaps in the future, parents might choose to pay the extra to make their children more radiation-resistant so they have a better chance of being selected as astronauts?

New developments will probably always be unsettling because they present people with a new way to look at themselves and the world around them but, after time, people accept these changes. Genetic screening will probably present a similar scenario. Fear will be present in the beginning but, as time passes, people will begin to accept genetic screening as a norm and will see it as a benefit to society, even if some of the effects are negative. Perhaps ECM astronauts won't live in the bleak future portrayed in *Gattaca*, but prospective crewmembers will only be able to hide from their DNA for so long before it catches up with them.

CLONING

Many misconceptions about human cloning stem from science fiction. Movies such as *Sleeper* and *Boys From Brazil* suggest only dictators will be cloned, while *Invasion of the Body Snatchers* suggests that on emerging from their pods, fully grown clones will have no emotions and will be murderous zombies. In fact, science fiction probably has a lot to answer for because people's fear of cloning is most likely at least partly based on how cloning is portrayed on the screen. For example, in *Bladerunner*, synthetic people – replicants – were produced that were identical to the humans except they had no empathy. Because of the generally negative popular views of human cloning derived from science fiction books and films, experts have rushed to reassure the public that a human clone would in no way be the same person or have any confusion about their identity. That's because the brain can't be cloned or duplicated from a DNA blueprint. They're pleased to point out that a clone of Mel Gibson would not be Mel Gibson. Well, that's a relief!

So, how would space agencies realistically exploit human cloning? Well, one possibility would be to employ clones as workers as depicted in the film *Moon* (Figure 7.6). Duncan Jones's *Moon* takes place deeper in the 21st century, when humans have finally returned to the Moon to mine helium-3 to solve an energy crisis on Earth. In charge of the Moon's mining operation is astronaut/engineer, Sam Bell (played by Sam Rockwell), who leads a lonely and isolated existence. He's near the end of a three-year deployment and is waiting to return to his wife and daughter. When he wakes up after an accident in one of the helium-3 harvesters, he returns to the base to find himself! The other version of Sam – Sam 2 – claims to be there for the same three-year contract, starting when Sam first began. Things don't add up.

```
INT. REC ROOM — NIGHT.

Sam 1 takes his old seat before the model. Sam 2 stands over
him.
                            SAM 2
    What about the other clones?
```

Figure 7.6 Sam Bell (played by Sam Rockwell) in *Moon*, is an employee contracted by Lunar Industries to extract helium-3 from lunar regolith. He is stationed for three years at a lunar base with only a robotic assistant named GERTY for company. During a rover excursion to retrieve canisters of helium-3, he crashes his rover, damaging a helium-3 harvester. While recovering from his injuries, Sam investigates the damaged harvester, where he finds someone barely alive in the crashed rover: himself! The two Sams struggle to come to grips with each other's existence, both believing the other to be a clone. After asking GERTY if he's a clone, the harvester-rescued Sam learns that he is and that the original Sam Bell's genetic material and memories have been harvested for the production of hundreds of Sam clones. Image courtesy: IMDB.

> SAM 1
>
> 'Other clones?' Sam 1 just stares back.
>
> SAM 2 (cont'd)
>
> Yeah, we might not be the first two to be woken up (indicating the model). You said that thing had already been started when you got here. Well, who started it? There might be other clones up here right now. Think about it. How did I get here so quickly after your crash? They didn't ship me in from Central, there wasn't time. I must have come from the base.
>
> SAM 1
>
> That's ridiculous. Impossible. Why would they do that?
>
> SAM 2
>
> I bet there's some kind of secret room —
>
> SAM 1
>
> (laughing)
>
> Secret room?
>
> SAM 2
>
> Yeah, secret room, why not?
>
> SAM 1
>
> (losing his cool)
>
> You're the one who's lost your mind! I've been here for three years. I know every inch of this base. I know how many dust fibres are between those wall panels over there — why would they do that?!
>
> SAM 2
>
> Look. It's a company, right? They have investors, shareholders — shit like that. What's cheaper? Spending time and money training new personnel or just have a couple of spares here to do the job. It's the far side of the Moon, Sam!
>
> Extract from the film *Moon*, by Nathan Parker,
> *www.script-o-rama.com*

Shocked to find he is a clone, there is more bad news for Sam, because there are an infinite number of Sams (Figure 7.7) ready to mine the Moon! What are two desperate clones to do?

Figure 7.7 Sam Bell and his clone, from *Moon*. Image courtesy: IMDB.

```
The two Sams can hardly believe their eyes: stretching back
maybe one hundred feet are rows and rows of NUMBERED DRAWERS —
like an epic morgue — and on the other side of the room,
directly facing the drawers, an equally epic line of fridges
full of food . . . .

Spooky as hell.

For a few moments Sam 1 and 2 are too stunned to speak. They
stare down the length of the chamber. It must be as long as the
base itself.

Sam 2 opens a drawer beside them. The clone inside is bare
chested. He appears to be wearing a HOSPITAL GOWN, identical
to the ones they wore in the infirmary.

Sam 1 MARVELS at how the clone is identical to himself in every
way, down to the minutest of details -- same hair, same skin
tone, same fingernails -- a few of the drawers have different
colored lights on next to them. Empty drawers. Sams who have
been and gone.
                          SAM 2
        Why are there so many of them?
```

Extract from the film *Moon*, by Nathan Parker, *www.script-o-rama.com*

The film offers up some very difficult moral questions. For example, who *will* do the drudge work when we return to the Moon, travel to Mars, or establish outposts on the moons of the outer planets? What *should* the rights of space-faring clones be? If Sam's clones have the same memories, personality, and humanity as the real Sam, can it possibly be anything other than monstrous to "retire" them at the end of their working life? The film leaves you wondering, showing everything from the clones' fractured and limited point of view. It certainly has the viewer thinking about what choices will be made in the future when the technology becomes available; would the space agencies play things out as they appear in the film? It's worth thinking about, and *Moon* is certainly worth watching.

CLONING ETHICS

> "*Mary Had A Little Lamb*
> Mary had a little lamb, its fleece was slightly gray
> It didn't have a father, just some borrowed DNA.
> It sort of had a mother, though the ovum was on loan,
> It was not so much a lambkin as a little lamby clone.
> And soon it had a fellow clone, and soon it had some more,
> They followed her to school one day, all cramming through the door.
> It made the children laugh and sing, the teachers found it droll,
> There were too many lamby clones, for Mary to control.
> No other could control the sheep, since the programs didn't vary,
> So the scientists resolved it all, by simply cloning Mary.
> But now they feel quite sheepish, those scientists unwary,
> One problem solved but what to do, with Mary, Mary, Mary."
>
> Anonymous post on the internet

Recent polls taken of Americans after Dolly's announcement showed that two out of every three people find human cloning to be morally unacceptable. Many fear the possibility of a diminished sense of identity and individuality, but, of course, there are other ethical questions that arise from this issue. For example, if we think a cloned human might be troubled by psychological problems that would make their life less satisfactory, then would it be best not to bring that human into existence? Also, how many cloned humans should there be? Of course, these questions are just the tip of the iceberg as far as the ethical issues involving human cloning are concerned. We don't have the answers to these questions and maybe we never will, but it's worth discussing some of the arguments for and against.

Offensive, disgusting, grotesque, and repulsive are the kinds of terms used by many to describe the prospect of human cloning. These are the universal feelings of many intellectuals, religious officials, scientists, believers, and non-believers, who are repulsed by the prospect of the mass production of human beings, the compromise of one's individuality, the narcissistic attitude of those who will clone themselves, and Man playing God. They believe cloning humans is ethically wrong because it fosters a reductionist rather than holistic view of human nature while treating people as means, not ends, and it creates a pressure to use this technology and make it a goal. They also believe human clones will not have a good quality of life because they would be soulless replicas of human beings that could be used as slaves. In short, the programmed reproduction of humans would be dehumanizing and change what it means to be human.

Proponents of human cloning argue cloned humans will be fully fledged human beings, indistinguishable in biological terms from other members of the species. These advocates who approach cloning with an open mind see only gains from human cloning. For example, clones could be useful in experimentation, since researchers would not have to correct for individual genetic differences. Also, someone may want to be cloned for spare parts (as suggested in the film *The Island*), and scientists or political authorities intent on improving human stock could use it to produce perfect children or to accelerate the evolutionary process. In fact, these supporters argue the potential benefits of human cloning may be so great that it would be a tragedy if ethics should lead to a Luddite rejection of cloning.

It is difficult to foresee what the future will hold for human cloning and no doubt ethics committees will continue to deliberate on the issues and governments will decide what to do. However, if cloning is found to have no effect on the health or lifespan of experimental animals, it would be reasonable to conclude that the same would hold true for human beings. If this happens, space agencies may decide that human cloning is ethically admissible and begin to clone radiation-resistant astronauts for jobs like Sam Rockwell's.

8

Robotic surgery and telemedicine

Exploration class missions (ECMs) to Mars, the outer planets, and beyond will require extended diagnostic and therapeutic medical capabilities, but these will be constrained by equipment and training limitations. It would be nice to embark a full surgical suite onboard an interplanetary spacecraft, but logistics will prevent that. Instead, it will be necessary to employ other strategies and techniques to ensure the health and well-being of those venturing beyond low Earth orbit (LEO). Since ECMs are at least a decade or so in the future, it's not surprising that the medical training for such a mission has yet to be defined. Today's astronauts orbiting onboard the International Space Station (ISS) receive medical training and are under the care of a non-physician crew medical officer (CMO), who receives 40–80 hr of basic medical and procedural training before the mission. The crew can also avail themselves of onboard medical and procedural checklists and audio communication with a flight surgeon at Mission Control in Houston, Texas.

Onboard ISS, a two-second delay is expected most of the time, but the communication lag during interplanetary missions will be measured in minutes and hours. Because of this time lag, ECM crewmembers who develop a surgical condition, or get injured or ill, will require novel treatment strategies that may not parallel treatment on Earth. For example, radiological capabilities will be limited due to power and weight limitations, so CMOs may need to rely on ultrasound as a diagnostic procedure and perhaps combine it with remote expert guidance and interpretation.

TELEMEDICINE

One rapidly developing application of clinical medicine that ECMs are sure to utilize is telemedicine (Table 8.1). Telemedicine (Figure 8.1) is a discipline in which medical information is transferred through interactive audiovisual media for the purpose of consulting and sometimes remote medical procedures. At its simplest, telemedicine may be as simple as two flight surgeons discussing a case over a satellite link, or as complex as conducting a real-time consultation between Mission Control and the

142   Robotic surgery and telemedicine

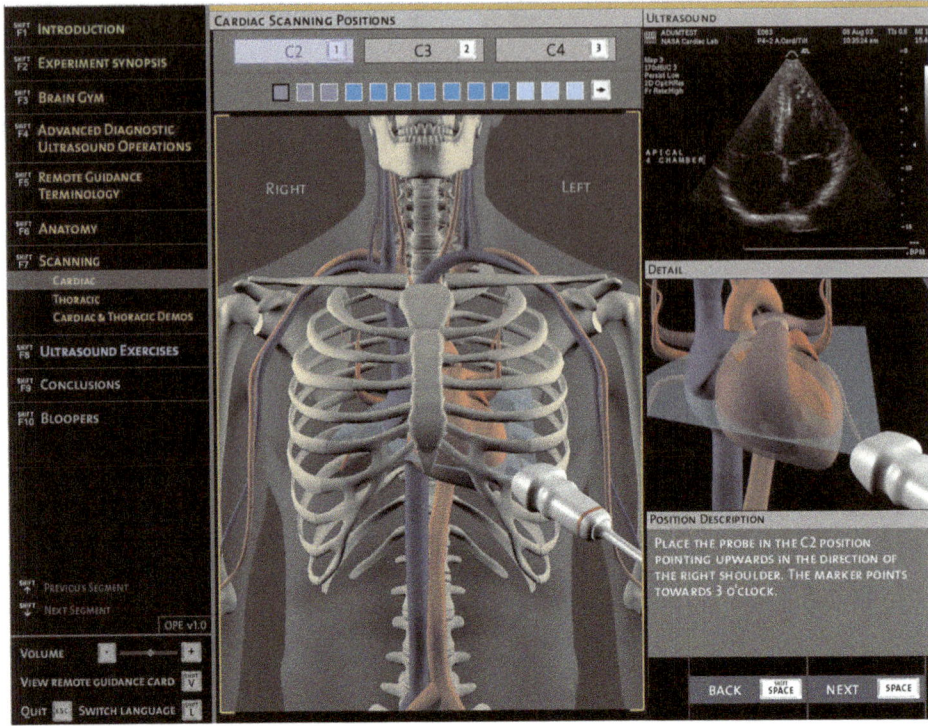

Figure 8.1 Telemedicine may require astronauts to use smart medical systems similar to the one depicted in this photo. The training session shown here highlights the basics of an ultrasound examination such as probe positioning, the location of the organ within the body, the size and structure of the organ, and what the correct ultrasound image should look like on the monitor. Image courtesy: Dr Scott Dulchavsky, NSBRI.

surface of the Moon. Telemedicine can be online (real-time) or offline, depending on the quality of the communication link. It can also be categorized based on the timing and the synchrony of the connection. For example, *store-and-forward telemedicine* means there is only one-way communication at a time. In this type of telemedicine, the CMO onboard the spacecraft would evaluate medical information offline and send it back to Mission Control following the evaluation. A second type of telemedicine is *remote monitoring*, in which flight surgeons collect information about patients from a distance. An example of this would be flight surgeons in Mission Control monitoring the physiological data of astronauts wearing bioinstrumentation, such as the Bio-Suit, which is discussed later in this chapter. The third category, *interactive telepresence*, provides real-time communication between two sites, which can be extended by various types of interactions. An example of this category of telemedicine would be a CMO in a base remotely directing the operation on a crewmember several hundred kilometers away. This leads us on to *telesurgery*, one of the applications of telemedicine that allows a surgeon to treat astronauts geographically separated by hundreds or thousands of kilometers. When the

Table 8.1. Concept of telemedical support.

Type of telemedical support	Features
Earth orbit and surface operations telesurgery	Viable within 380,000 km
	Less than 2-sec signal delay
	Appropriate for surface operations
Telementoring	Viable within 10,000,000 km
	Less than 50–70-sec signal delay
	Permanent video contact with Mission Control
Consultancy telemedicine	Within the range of Mars
	Up to 44-min signal delay
	Preoperative simulations and consulting

connection is not reliable, a surgeon might use *telementoring*, a technique using almost real-time video and audio from the operating room. In this application, a virtual classroom is created, permitting the surgeon, who remains in their hospital, to instruct a novice in a remote location how to perform a new operation or use a new surgical technology. When communication is compromised, or the time lag doesn't allow real-time connectivity, *consultancy telemedicine* (sometimes called telehealth consultancy) can be used – an application that only requires limited access to the remote site.

An integral element of telemedicine and its applications is telerobotics. While robot surgery may sound futuristic, the robotic uprising has been headed for the operating room for quite a while. Surgical robots can already image your body in three dimensions, pinpoint the precise location of a buried cyst, and direct mechanical arms to perform biopsies. In fact, surgical robots have lent surgeons a helping hand for years; in the early 1990s, surgeons used robots to drill holes for hip replacements, and in 2001, a robot named Zeus translated the movements of a surgeon in New York to an operating room in France. In 2007 alone, surgeons performed some 85,000 procedures with just one type of surgical robot, enabling them to work with remote arms through small openings, using specialized software to reduce hand tremors.

It's fitting that robots will assume an important role in space medical operations because they have already played a significant role in space exploration, enabling us to see places that humans have yet to visit. Often, when we talk about robots doing the tasks humans do, we're talking about the future, but robotic surgery is already a reality. As this book is being written, doctors around the world are using sophisticated robots to perform all sorts of surgical procedures on patients, using techniques that will one day be used not only during the transit to and from the planets, but also on the surface. Today's surgical robots can be classed into three systems: supervisory-controlled systems, telesurgical systems, and autonomous systems. The main difference between each system is how involved the human is when performing a surgical procedure. At one end of the scale, doctors perform surgery with the assistance of a robot, while at the other end of the scale, robots perform surgery without the direct intervention of a surgeon. While many people

144   **Robotic surgery and telemedicine**

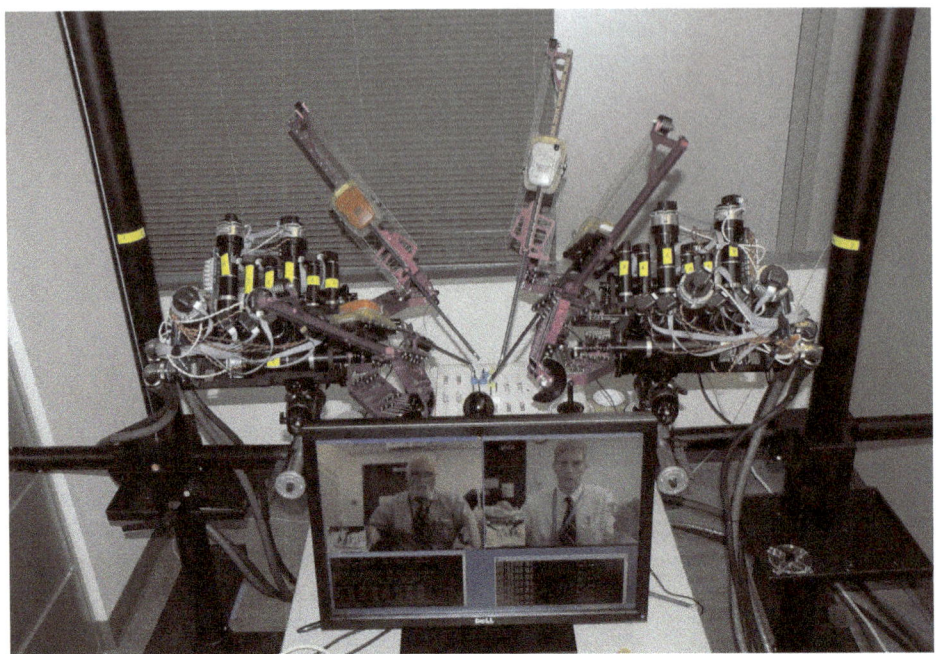

Figure 8.2 Dr Mika Sinanan (UW, Department of Surgery) and Dr Thomas Lendvay collaboratively teleoperating Raven IV located at the Bionicas Lab at UCSC (Santa Cruz, CA) from the UW in Seattle, WA. Image courtesy: UCSC.

might feel uncomfortable handing surgery over to a robot, astronauts bound for the outer planets will have no choice, so it's worth discussing how these systems might work.

SUPERVISORY-CONTROLLED SYSTEMS

First, we'll discuss supervisory-controlled systems (Figure 8.2). These are robots that can only act under the direction of a surgeon. In 2000, the US Food and Drug Administration (FDA) approved the da Vinci Surgical System for laparoscopic[1] procedures, making it the first robotic system allowed in American operating rooms. The da Vinci, which comprises a viewing and control console together with a surgical arm, uses technology that allows the human surgeon to get closer to the surgical site than human vision will allow, and work at a smaller scale than conventional surgery permits. It may look cumbersome, but by the time ECMs are possible, a scaled-down version will be available.

The ECM version of the da Vinci robot will be used mostly as a diagnostic tool

---

[1] Laparoscopic surgery, also called keyhole surgery, is a modern surgical technique in which operations in the abdomen are performed through small incisions (0.5–1.5 cm).

and to perform surgical procedures to examine the abdominal and pelvic organs, or the thorax, head, or neck. Tissue samples could also be collected for biopsy using the robot and even malignancies could be treated when combined with other therapies. The CMO using the robot would make three or four incisions no larger than the diameter of a pencil in the patient's abdomen. This would allow the surgeon to insert the stainless-steel rods you can see in the photo (the robotic arms hold the rods in place). One of the rods has two endoscopic cameras that provide a stereoscopic image, while the other rods have surgical instruments the CMO would use to manipulate and suture tissue. Unlike in conventional surgery, the CMO wouldn't touch these surgical instruments; instead, he/she would sit at the control console near the operating table and look into a viewfinder to examine the stereoscopic images provided by the endoscopic camera inside the patient. The images show the surgical site and the surgical instruments mounted on the tips of the surgical rods. The CMO would use controls similar to a joystick located underneath the screen to manipulate the surgical instruments. Each time the CMO moved one of the joysticks, a computer would send a signal to one of the instruments, which would move in tandem with the movements of the CMO's hands. Once the surgery is complete, the CMO would remove the rods from the patient's body and close the incisions. One problem with these robots is that there is no room for error – these robots can't make adjustments in real time if something goes wrong, so surgeons have to watch over them and be ready to intervene if something goes awry.

TELESURGERY

Another system that will be in the ECM medical toolkit is the telesurgical robot. *Telesurgery* is a procedure that involves a doctor performing delicate surgery miles away from the patient. In this type of surgery, the doctor controls the robotic arms from a computer station at a remote location. For example, it's possible for a flight surgeon in Houston to operate on a patient onboard the ISS, or a doctor in Washington to operate on a patient in an underwater habitat off the coast of Florida (Figure 8.3). While the technology provides people in remote environments with life-saving emergency medical care if no doctor is available, there are some problems, one of which is latency. Latency is the time delay between the surgeon's moving his/her hands to the robotic arms responding to those movements. Although surgeons can train to overcome this latency, the system only really works for crews in LEO or astronauts on the surface of the Moon, where the lag time is about a second or two. While such a system wouldn't be used in transit, for surface operations, it would be ideal.

Telerobotic surgery was recently conducted during NASA's 12th Extreme Environmental Mission Operations (NEEMO) mission in 2007. The NEEMO 12 team, comprising NASA astronauts, surgeons, and professional divers, conducted the surgery onboard the Aquarius undersea laboratory (Figure 8.4), which rests 18 m below the ocean's surface off the coast of Key Largo in the Florida Keys. Joining commander, Heidemarie Stefanyshyn-Piper, on the Aquarius mission were

Figure 8.3 NASA Extreme Environment Mission Operations (NEEMO). A NEEMO 12 crewmember peers through a window port, where the University of Washington surgical robot known as Raven is visible inside the Aquarius undersea laboratory during NASA's NEEMO 12 mission in May 2007. Image courtesy: NASA.

Figure 8.4 Aquarius, which is 14 m long and 3 m wide, contains about the same habitable area as the International Space Station's (ISS) Russian-built Zvezda service module, which serves as the primary living quarters for ISS crews. Image courtesy: National Oceanic and Atmospheric Administration.

fellow NASA astronaut Jose Hernandez, flight surgeon Josef Schmid, and University of Cincinnati researcher Tim Broderick, who watched over telerobotic surgery experiments performed by a telesurgical robot known as Raven (Figure 8.5) – a portable two-armed automaton built by researchers at the University of Washington in Seattle.

Thanks to the robot's modular construction, crewmembers were able to transport the surgical robot to Aquarius in dive bags. Once inside the habitat, the crew pulled out the instruction manual and reassembled the robot before hooking up with researchers at the University of Washington's BioRobotics Lab in Seattle. From their conference room on the west coast, the researchers were able to guide the robot through a series of tasks on a simulated patient using a commercial internet connection and by transmitting the signals via a wireless connection to a buoy on the habitat, which was, in turn, routed through a hardwired umbilical into Aquarius. The robots performed several simple tasks such as suturing and some laparoscopic techniques.

148 **Robotic surgery and telemedicine**

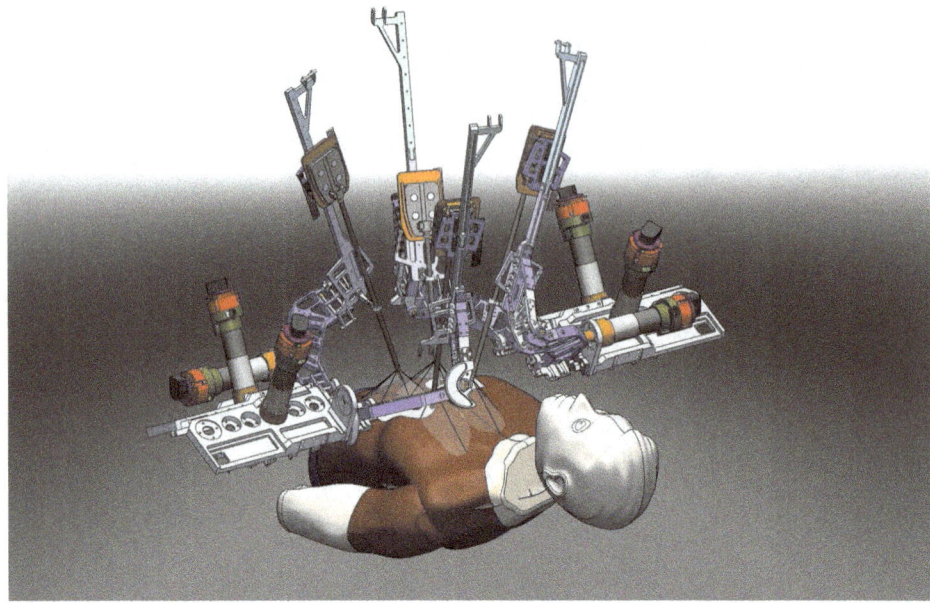

Figure 8.5 To reproduce the dynamics of two surgeons interacting with the surgical site using a robotic system, four dexterous arms along with two pairs of eyes are needed. Raven IV (pictured) is a surgical robotics system that was developed at the University of California. The system facilitates a collaborative effort of two surgeons interacting with the surgical site in teleportation. Each surgical arm is based on a spherical mechanism with a remote center located at the point of entry of the tool into the human body. The system architecture allows two surgeons in two remote locations to connect via an internet connection. Image courtesy: UCSC.

AUTONOMOUS

The telesurgery techniques performed by the NEEMO crew will no doubt be used on planetary surfaces. Astronauts on the surface may get injured at a location several hundred kilometers from the base, which would require the help of the CMO, who could perform the medical procedure remotely. However, while telemedicine can provide diagnosis and treatment in near real time, what happens if there is a problem en route when round-trip communications might be measured in several minutes or hours? Well, the most likely option will be some sort of automated system. For example, imagine a crewmember suffering from chest pain 40 million kilometers from Earth. The CMO would examine the stricken astronaut and enter the results into a computer equipped with artificial intelligence (AI). The AI would interpret the information and generate a diagnosis. Here's how it might work:

The AI asks a question, and the answer dictates the next question.

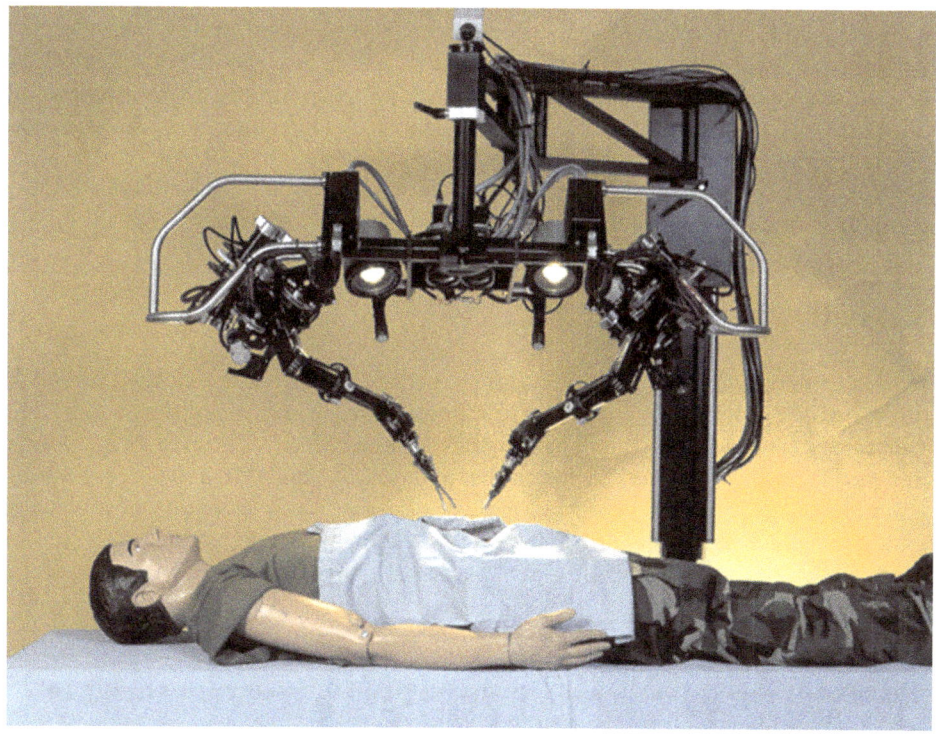

Figure 8.6 Raven. Image courtesy: UCSC.

AI: *What seems to be the problem?*
Stricken astronaut: *My chest. It's my chest. Damn, it hurts.*
AI: *What sort of pain is it?*
Stricken astronaut: *Pressure pain, like an elephant's sitting on me.*
AI: *Do you have pain in your arms and shoulders?*
Stricken astronaut: *Yes.*
AI: *On a scale of 1 to 10, with 10 being the worst pain you've ever experienced, how much pain do you have?*
Stricken astronaut: *Nine.*
AI: *When did the pain start?*

And so it goes. To science fiction fans, the development of AI is just as appealing as it is to space agencies tasked with providing autonomous health care for its astronauts. While the notion of having a robot surgeon (Figure 8.6) operate on an astronaut in a remote location with no human guidance may seem far-fetched, this next step in human–robot relations is already being developed by a team of biomedical engineers at Duke University. The engineers have already demonstrated that a robot, without any human assistance, can locate a lesion in simulated human organs, guide a device to the lesion, and take multiple samples of the lesion. To conduct their tests, the Duke team used raw turkey because the tissue closely resembles that of humans in texture and density, and appears similar when scanned by ultrasound. An existing robot arm with an ultrasound system was used to perform the searching, with the ultrasound serving as the robot's "eyes" by collecting data from its scan and locating its target. The robot was "controlled" by an AI program that took the real-time 3-D information, processed it, and gave the robot specific commands to perform. At the pace at which the technology is being developed, it won't be long before autonomous robots will be performing more complex surgical tasks. And they won't just be performing from the outside.

NANOBOTS

Nanotechnology involves manipulating properties and structures at the nanoscale, often involving dimensions just tiny fractions of the width of a human hair. The technology is already being used as the basis for more effective drug delivery systems and is in the early stages of development as scaffolding in nerve regeneration research. In the future, nanotechnology will also aid in the formation of molecular systems that will be similar to living systems and that could be the basis for the regeneration/replacement of body parts. As usual, nanotech is a subject Hollywood has tackled:

```
INT. THE LEFT VENTRICLE
```
Jack's fear causes a sudden rush of blood into the ventricle, followed by strong ventricle contractions. The Pod is buffeted and tossed like a beer can in the pounding surf.
```
INT. THE POD
```
Tuck fights for control of his craft. It shakes and vibrates. Interior lights dim and flash. It seems the Pod will tear apart at the seams. Tuck almost blacks out. Then he sees:
```
THE AORTIC ARCH
```
Offering four distinct pathways out of the heart.
```
TUCK — INSIDE THE POD
                        TUCK
```
    The aortic arch! I'm almost out! I've got to make it through the opening on the right!

He pulls hard on the control stick. The Pod begins to turn. It trembles and shakes. Portions of the sidewalls are pushed inward by the tremendous pressure. Gauges shatter. Warning lights flash and BUZZ.
```
THE POD
```
SLAMMING against the muscle-lined vascular wall. Bouncing back. Shooting through the proper 'archway' into the relative calm of the AORTA.

>   TUCK (V.O.)
>
>   I'm through!
>
> THE BLACK POD
>
> Igoe views his sonar scanner in disbelief. The 'Blip' tells
> him that Tuck has made it safely through the heart.
>
>   IGOE
>
>   Impossible!
>
> INT. THE STOMACH
>
> Tuck's Pod BLASTS through the membrane lining of the stomach
> wall.
>
>   TUCK (V.O.)
>
>   I'm in the stomach. No sign of the bandit.
>
> INT. THE BLACK POD
>
> Igoe enters the stomach. He can see Tuck's Pod up ahead. He
> swings his Laser Gun Sighting Device into place.

Excerpt from the screenplay *Innerspace*,
by Jeffrey Boam, *www.dailyscript.com*

In the classic science fiction film *Innerspace*, down-on-his-luck naval aviator Tuck Pendleton (played by Dennis Quaid) is selected to participate in an experimental project that places him in a miniaturized submersible pod injected into a rabbit. After being miniaturized, the experiment turns bad when the lab is attacked. The experiment supervisor escapes with the miniaturized Tuck in a syringe. After sustaining a fatal gunshot wound, he injects Tuck and the pod into hypochondriac supermarket clerk Jack Putter (played by Martin Short). After establishing contact with his new host, Tuck must figure out how to get out of Jack before he runs out of oxygen. After contacting the lab, Tuck and Jack are informed that there's another group of scientists attempting to use miniaturization for espionage. The attack on the lab was an attempt to steal a chip vital to the process. Having secured the chip, villains miniaturize Mr Igoe and send him into the body of Jack, to extract a second chip required for re-enlargement installed in Tuck's pod.

In many ways, *Innerspace*, made in 1987, foreshadowed many of the aspects of medical nanotechnology. While it is unlikely humans will be miniaturized and injected into bodies, a new class of tiny robots has been developed capable of fending off disease, performing surgery, and even performing onsite repairs. For example, *wormbots* are capable of squeezing through the digestive system by switching between *undulatory* motion to wiggle like a snake and *peristaltic* locomotion to shrink and elongate like a worm. Other micro-robots use tail-like *flagella* and *cilia* to swim through the body. More advanced nanobots will be able to sense and adapt to environmental stimuli such as heat, light, sounds, surface textures, and chemicals, perform complex calculations, move, communicate, and work together, and even replicate themselves. *Nanoparticles* will attach to specific cells and provide medical images of their location and structure, hollow *nanocapsules* loaded with pharmaceutical contents will deliver payloads to sites around the body, while *nanomedibots* will repair vital tissue damaged by injury.

Cell repair machines will be similar in size to bacteria, but their compact parts will allow them to be more complex. They will travel through the circulatory system, and once inside a cell, the repair machine will assess the situation by examining the cell's contents and activity, before taking action. Among other things, these repair machines, guided by *nanocomputers*, will be able to recognize and repair DNA damage (remember the vasculoid in Chapter 4?). They will be so small that one machine will fit in 1/1,000th of the volume of a typical cell, yet it will hold more information than the cell's DNA! Working molecule by molecule and structure by structure, these repair machines will repair whole cells and, if necessary, entire organs. Think how beneficial these machines will be to radiation-ravaged astronauts living on the surface of Callisto or orbiting other radiation hotspots such as Europa. As the radiation inflicts damage on DNA material, fleets of repair machines will simply be deployed to repair the cross-links in the DNA and correct any mutations.

TRAUMA PODS

Imagine an astronaut wounded several hundred kilometers away from base. A micrometeorite has penetrated his suit and passed through his leg. He's bleeding from the femoral artery, his blood pressure is plummeting, and he's slipped into unconsciousness. In most circumstances, he's as good as dead, but this guy is lucky. Rumbling onto the scene is an unmanned mobile intensive care unit (ICU), which pulls up alongside him. A hatch opens (Figure 8.7), a robotic stretcher scoops him up into its armored belly, and starts heading for home. Inside, a robotic surgeon, controlled by the CMO at the base, monitors the astronaut as information is uploaded from sensors in the astronaut's Bio-Suit. The stricken astronaut is given oxygen, his vitals are assessed, and the robot begins stabilizing life-threatening injuries while sending data to the CMO. Just minutes into the autonomous evacuation, the robot scans the astronaut with a computed tomography (CT) system and sends the information to the CMO. Based on the CT data, the robot gets to work (Figure 8.8). Using tourniquets, gauze, and other supplies dispensed from an automated medical kit, the robot stems the flow of blood and laser-welds the wound together. Fifty kilometers from the base, the astronaut's vital signs stabilize, and he regains consciousness. A few minutes later, the mobile ICU picks up a transmission from the base shuttle, which is en route to pick up the astronaut and bring him back to base. The mobile ICU deploys a gurney that automatically rotates to an egress port on top of the vehicle. The shuttle (automated, of course) makes a low pass, hovers above the mobile ICU, hooks up the gurney, and pulls the grateful astronaut inside. Ten minutes later, the injured astronaut is making a recovery inside the base, attended to by the CMO. Meanwhile, the mobile ICU returns to its patrol duties, waiting to respond to any other events.

Figure 8.7 Conceptual image of the Trauma Pod being lifted into the mobile unit. The Trauma Pod is a groundbreaking unmanned medical treatment system designed to stabilize injured soldiers within minutes after a battlefield trauma and administer life-saving medical and surgical care prior to evacuation and during transport. Image courtesy: SRI International.

Science fiction?[2] Actually, the mobile ICU depicted in the story is based firmly in reality. The mobile ICU is known as the Trauma Pod (TP), which has evolved from a working project of the US Department of Defense that's well under way and the final version is expected to come online in the 2015–2020 timeframe. It's the creation of the US government's Defense Advanced Research Projects Agency (DARPA) Defense Sciences Office (DSO), and is being developed by a consortium of firms led by SRI International of Menlo Park, California – a developer of telesurgery systems mentioned earlier. While the TP is designed for military use, it's easy to envisage an ECM version of the system. Here's a more detailed description of how it would work.

The space version of the TP would be designed to provide urgent medical care on site to astronauts with serious injuries on planetary surfaces. Given the nature of these environments, the TP would be required to provide specialized medical care and also deal with complications caused by the evacuation process. In a nutshell, the space TP (STP) would need to be a very versatile system, capable of deploying

---

[2] The concept has been mentioned in science fiction. In his 1972 novel, *The Godmakers*, Frank Herbert wrote about crèche pods, which were fully automated trauma pods.

154   Robotic surgery and telemedicine

Figure 8.8 Trauma Pod conceptual image: surgical tools are poised for operation. Image courtesy: SRI International.

autonomously, performing diagnostics, and performing acute life-saving interventions. Achieving this versatility will obviously require several subsystems that would need to be integrated with a layer of well defined interfaces.

At its simplest, the STP would comprise two cells: a control cell, where the surgery is monitored by an administrator AI, and a surgical cell, where the surgery is performed by the medical AI. The control cell would be located at the base, while the surgical cell would be deployed on an autonomous rover. To illustrate the various systems, let's use the example of the injured astronaut again. We'll assume the rover has loaded the astronaut onboard and is assessing the damage. Before performing any surgery, the STP patient registration subsystem (PRS) would hold the astronaut on the platform and scan him stereoscopically, creating an external 3-D model allowing the STP robots to move safely around the astronaut. Then, the STP would generate a medical encounter record (MER) containing the astronaut's medical history, triage information, preoperative diagnosis, and surgical notes. The next step would be to conduct a CT scan of the astronaut using a patient imaging system (PIS), consisting of a tube mounted on an overhead rail that slides across the astronaut. The scan, which would take about 15 sec, would generate a 3-D image of the astronaut's internal organs. Once it had decided what to do, the surgical robot subsystem (SRS) would go to work. Serving the SRS would be the scrub nurse subsystem (SNS), which would deliver supplies to the SRS and exchange tools during the operation. The SNS would work together with the supply dispensing subsystem (SDS), which would store, de-package, dispense, and discard consumable supplies. The SDS would be handed the tools by an automated tool rack, which would be capable of sterilizing used tools and adjusting the tension and orientation of the tools if required. Supervising the sequence of events would be an AI, coordinating the actions and the various subsystems and prioritizing tasks and responsibilities. By using clinical protocols and procedures, the AI would also be responsible for monitoring safety and making sure the surgery proceeded according to plan. Event times, surgical procedures (incision, debridement, placement of stents, administration of fluids, and use of instruments), and clinical protocols would all be recorded by the AI, ready to be transferred to the main medical suite at the base.

The scenario might sound far-fetched, but Phase 1 studies have already been conducted by SRI International. The studies, conducted in 2008, verified the feasibility of conducting a robotic surgical operation with no medical personnel on site.

MEDICAL MONITORING

In addition to STPs and a new suite of surgical tools, ECMs will also demand a recalibration of astronaut health monitoring. This will be required due to the length of ECM surface missions that will feature astronauts conducting lengthy extra-vehicular activities (EVAs). The system that future planetary explorers might use is the Bio-Suit (Figure 8.9) – a system developed by Professor Dava Newman and her team from the Massachusetts Institute of Technology (MIT) in collaboration with the NASA Institute for Advanced Concepts (NIAC).

Based on the concept of biomechanically and cybernetically augmenting human performance, the Bio-Suit is envisioned to function as a second skin by providing mechanical counter-pressure (MCP). Whereas today's EVA suits weigh at least 40 kg, Newman's skin-tight alternative weighs as little as 10 kg! Constructed of spandex and nylon, the multi-layered suit hugs the body's contours, and its MCP technology ensures constant pressure is applied to the surface of the body. This pressure is needed not only to counteract the vacuum of the surface environment and to maintain the body's homeostasis, but also to avoid blood pooling. Maintaining an even pressure over the surface of the human body has, until now, been achieved by utilizing the bulky gas-pressurization systems embodied by traditional EVA suits, but thanks to new MCP technology that works along lines of non-extension (those lines along the body undergoing little stretching as the body moves), a gas-pressurization system is no longer necessary.

Another advantage of the Bio-Suit is its suite of sensors that monitor biomedical signals such as heart function, oxygen consumption, and body temperature. Another suite of biochemical sensors provides information concerning body fluids, and dosimeters measure the local radiation environment. The sensors acquire physiological data continuously and the astronaut can view the parameters in a stereoscopic image. Another neat feature is the thermal sensor network that reacts to fluctuations in environmental conditions, which ensures the astronaut is

Figure 8.9 The revolutionary Bio-Suit. Image courtesy: Professor Dava Newman, MIT: Inventor, Science and Engineering; Guillermo Trotti, A.I.A., Trotti & Associates, Inc. (Cambridge, MA): Design; Dainese (Vicenza, Italy): Fabrication; Douglas Sonders: Photography.

comfortable despite temperature fluctuations. It's an ideal suit for ECM surface EVAs, which will last six to eight hours and require crewmembers to visit isolated locations and perform complex activities; if an accident occurs, having access to the information provided by a wearable sensor system would greatly improve the chances of survival.

The Bio-Suit, surgical robots, and the spectrum of telemedicine technologies are important enabling technologies for ECMs. Hardware such as the TP and nanotech are technology accelerators that drive the development of more capable devices that ECM astronauts will require for delivering advanced medical care in remote locations such as the surface of Callisto and destinations beyond. If you're having trouble imagining the concept of devices that can execute complex surgery autonomously, remember that anesthesia was viewed as pretty radical for its day, too.

9

Stasis

```
INT. HYPER-SLEEP CHAMBER

A RUMBLE passes . . . EYELIDS TWITCH . . . A GASP.

Inside a coffin-sized chamber, a dim glow of light seeps
through a small glass porthole. A MAN is asleep, a mask
attached to his face, tubes of liquid feeding into his arms.

CORPORAL BOWER, military physique, chiseled features. His
chest lightly expanding, blood pulsating through his veins.

Lights FLICKER PAST and his body comes to life, muscles
flinching, breath quickening. He squirms with discomfort,
ripping away the mask. He gasps, trying to SCREAM.

His scream is muffled under three inches of glass. The hatch
door is hit with a THUD and white gases erupt from the edges.

THUD, THUD THUD! The door bursts open and BOWER scampers out,
struggling to rip away the feeding tubes. He HITS the floor,
naked and gasping. His pores ignite, dripping sweat. He tries
to get to his feet, but his legs wobble underneath him,
collapsing. He yells at the floor.
```

Excerpt from the screenplay *Pandorum*,
by Travis Milloy, *www.mypdfscripts.com*

The above excerpt is taken from one of the opening scenes of Christian Alvart's excellent 2009 science fiction film *Pandorum*. In *Pandorum*, Bower (played by Ben Foster) wakes up from hypersleep to find himself alone, with no memory of who he is, what he is doing, or what happened to the crew of the 60,000-passenger spaceship, *Elysium*. He proceeds to wake up Payton (played by Dennis Quaid), who is also suffering from memory loss. They are unable to access the flight deck and can't communicate with other members of the crew, including the flight crew team, whom they are supposed to relieve.

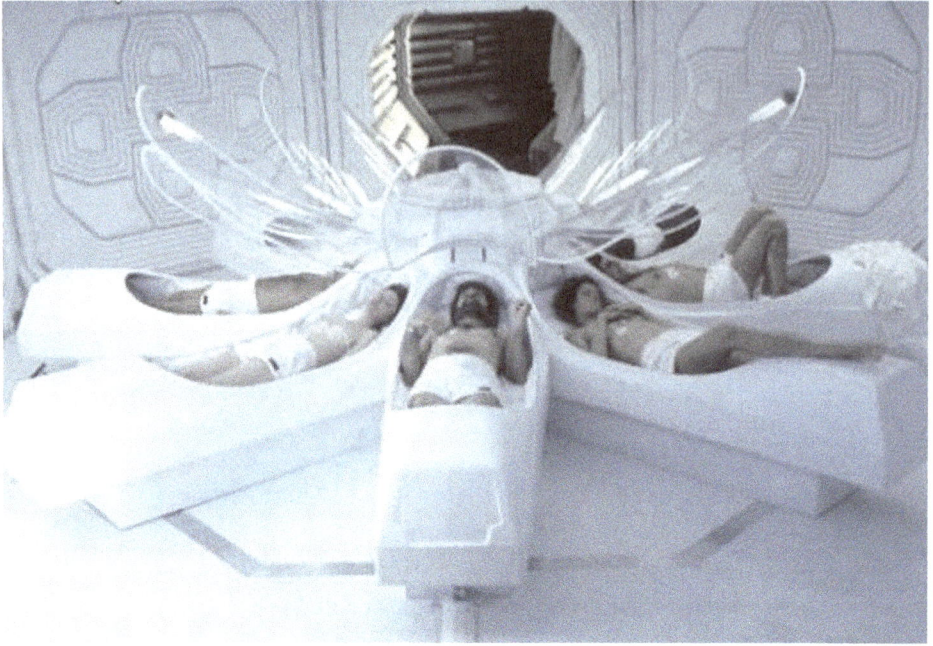

Figure 9.1 Hypersleep, Hollywood style. The crew of the *Nostromo* lies sleeping in their hibernaculum in the classic science fiction movie, *Alien*. Image courtesy: IMDB.

Stasis, or hypersleep (Figure 9.1), is a popular science fiction concept[1] akin to human hibernation or suspended animation, but while suspended animation often refers to a greatly reduced state of life processes, *stasis* implies a complete cessation of these processes, which can be restarted when stasis is removed. Now, you may be wondering why anyone would take the risk of being put to sleep for several months or years. Well, there are a number of reasons. Let's consider the living conditions first.

First of all, ask yourself this question: could you handle several years in space with a crew of only four of five? Bear in mind that you'd be doing *everything* with them. Eating, sleeping, working, waiting, occasionally responding to emergencies, followed by more eating, sleeping... well, you get the picture. Take a quick glance around your workplace and imagine spending time with your workmates 24 hr a day, seven days a week for three years or more. It would be enough to drive anyone mad. Even close-knit families find it difficult to get along in close quarters, and don't forget, during an exploration class mission (ECM), it's not as if you can simply walk

---

[1] In the science fiction film *Alien* and its sequel *Aliens*, crewmembers hypersleep their way to other planets. At the beginning of *Aliens*, Ripley has been in stasis for 57 years as she drifted in her "lifeboat" after the events of *Alien*. Another notable use of stasis is in the classic *Red Dwarf* television series, where a stasis chamber is used to preserve Dave Lister for 3 million years.

Table 9.1. Effect of stasis on the life-support-system requirements.

Life-support area	Purpose	Effect of stasis
Atmosphere management	Atmosphere control, temperature/humidity control, atmosphere regeneration, ventilation	Reduced heating and regeneration requirement
Water management	Provision of potable and hygienic water, recovery, and processing waste water	Reduced dramatically
Food production/storage	Provision and production of food	Reduced dramatically
Waste management	Collection, storage, and processing of human waste and refuse	Reduced dramatically
Crew safety	Fire detection and suppression	Increased
Crew psychology	Maintenance of crew mental health	Reduced dramatically
Crew health	Bone demineralization and muscle atrophy	Augmented systems required

out of the door and escape – even polar explorers had that option; you're committed for the duration of the mission. But what about today's astronauts? After all, they seem to get on fine during their six-month stints onboard the International Space Station (ISS). Well, the ISS is very different from an interplanetary spaceship. For one thing, the workload is very busy, which means the crew just don't have much time to think about the annoying habits of their crewmates. Second, the ISS is a big orbiting facility, so if crewmembers *do* need some downtime, it's not too difficult to find an empty module to chill out in. In contrast, an interplanetary spacecraft will be a cramped space, perhaps no larger than a school bus. Finally, the ISS has several changes of crew during a typical six-month mission, which means you get to see new faces once in a while. Not so on a trip to Callisto.

The next reason for putting astronauts into stasis is a simple question of logistics. When humans venture beyond Mars to far-flung destinations such as the outer planets, having some sort of stasis capability will be mandatory just to avoid having to lug along tonnes of life-support supplies. Even with a Variable Specific Impulse Magnetoplasma Rocket (VASIMR) capable of reducing transit times to Callisto to just 847 days, such journeys (Figure 9.2) will require huge life-support-related masses. Even for a short two-year round-trip mission to Mars, the Equivalent System Mass (ESM – a measure taking into account the quantity of consumables and the equipment required to maintain/deliver/manage it) for food alone is 30 tonnes! Then you have the weight of water, atmosphere provision, and waste management to consider. And what about the mass penalty of the exercise machines and recreational facilities? It's just not reasonable. Because of all this weight, mission planners will have to consider an alternative approach that drastically reduces the requirements (Table 9.1) imposed on the life-support system (LSS) in virtually every area. Fortunately, the answer can be found in the natural world: hibernation.

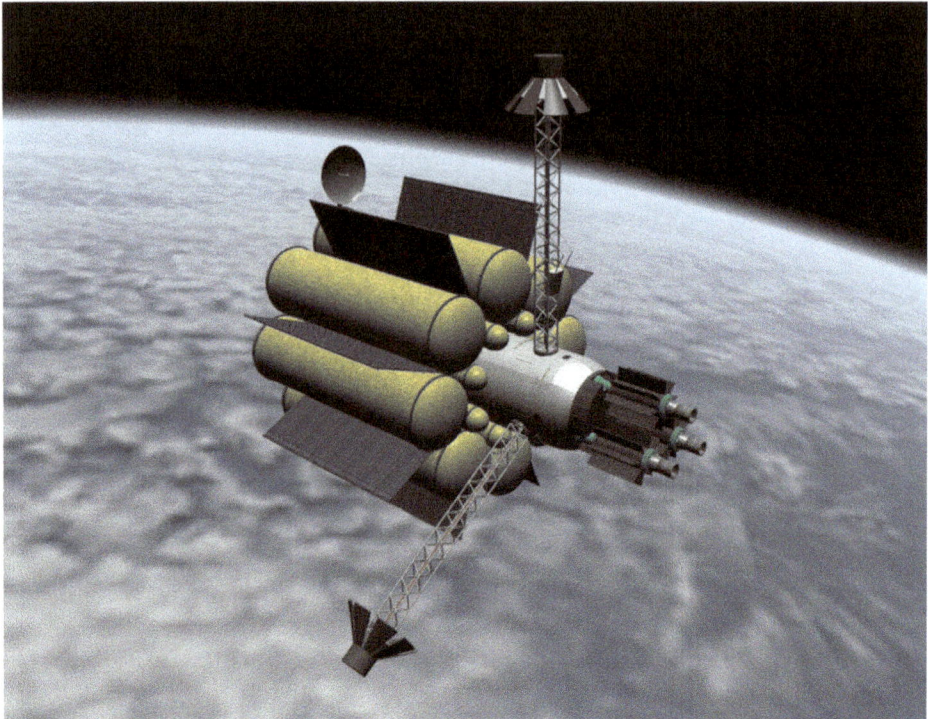

Figure 9.2 The Variable Specific Impulse Magnetoplasma Rocket (VASIMR) uses an electromagnetic thruster for spacecraft propulsion. Former astronaut Franklin Chang-Diaz created the VASIMR concept and has been working on its development since 1977. Image courtesy: NASA.

ANIMAL HIBERNATION

In nature, hibernation is a time when animals "sleep" through cold weather, but this sleep isn't like human sleep, in which loud noises can wake you up. In true hibernation, animals can be moved around or touched and not know it, although you probably wouldn't want to test this theory with a bear! In fact, hibernation is just one of five forms of dormancy displayed in animals, the other four being sleep, torpor, winter sleep, and summer sleep. To prepare for hibernation, animals eat more food than usual in the fall, to store fat needed to survive the period of hibernation. Some animals, such as the black bear (Figure 9.3), also store food in caches, while some species employ both methods. Generally, food caches are used by true hibernators, while winter sleepers tend to rely more on accumulating fat reserves. The pre-hibernation fattening period results in white adipose tissue being stored, although some animals also lay down reserves of brown fat, which accumulates in patches along the neck and major blood vessels.

Figure 9.3 Bears are often thought of as hibernators, although bears do not go into true hibernation because during a bear's winter sleep state, the degree of metabolic depression is much less than that observed in smaller mammals; the bear's body temperature remains relatively stable (depressed from 37°C to approximately 31°C) so it can be easily aroused. Image courtesy: Wikimedia.

After packing on the pounds, hibernators search for a place to hibernate. In hibernation parlance, this place is termed the *hibernaculum* and it can be anything from a cave to a hole in a tree. The time of entering hibernation varies among animals; some, like the alpine marmot, hole up in late September, while others go to sleep later in the year. Scientists aren't sure how the time to start hibernation is determined, but it is thought that animals rely on certain cues such as the length of day. It is also thought that some animals enter hibernation as a result of the action of a "trigger molecule" that initiates hibernation. One such molecule has been found in arctic ground squirrels (Figure 9.4) and has been termed the Hibernation Induction Trigger (HIT). The HIT's action is not completely understood, but it is likely a similar technique will be used for humans.

Once a hibernating animal enters hibernation, a number of things happen. In the ground squirrel (Panel 9.1), respiratory rate drops from as high as 200 breaths per minute to as low as four to five breaths per minute and the heart rate falls from 150 to five beats per minute (Table 9.2). The precipitous drop in breathing and heart rate is part of the overall reduction in metabolic rate. Other changes include a fall in

Figure 9.4 The arctic ground squirrel hibernates over winter from early September to late April, during which it can reduce its body temperature from 37°C to as little as −3°C. Image courtesy: Wikimedia.

body temperature, with some animals such as the arctic ground squirrel cooling to below freezing! But the change in metabolic rate doesn't stay the same throughout the hibernation period because hibernating animals occasionally wake to eat, drink, and eliminate wastes. During these wakeful periods, the body temperature and other physiological parameters return to normal levels. In contrast, the winter sleepers (bears, which many people think of as the classic hibernating animals, are actually just deep sleepers) stay dormant throughout the hibernation period without eating or drinking.

During the hibernation period, animals use 70–100 times less energy than when active, allowing them to survive until food is once again plentiful. At the cellular level, animals get their energy in the form of adenosine triphosphate (ATP), which is produced in the mitochondria. A chemical process occurs inside the cell which supplies the energy required for maintaining basic physiological function during the hibernation period. Once the animal exits hibernation, the biochemistry and metabolism return to normal, although the animal may not feel 100%; as you can imagine, waking up after spending several months asleep can be a little discombobulating!

Panel 9.1. How squirrels hibernate

Richardson's ground squirrels, also called gophers, hibernate for four to nine months of the year, depending on age and gender. Each animal hibernates underground by itself in its own hibernaculum. The squirrels spend 85–92% of hibernation in the physiological state of torpor, during which time their body temperature is about the same as the temperature of the surrounding soil, and heart rate, respiration, and metabolism slow dramatically. In January, these squirrels spend 20–25 consecutive days in torpor, with their body temperature dropping as low as 0°C. In between the periods of torpor, the squirrels re-warm to the normal mammalian body temperature of 37°C. The revivals last less than 24 hr and consist of a two-to-three-hour re-warming period, followed by 12–15 hr when the animal is warm but mostly inactive. Body temperature then slowly cools back down to ambient soil temperature and the squirrel enters another period of torpor. Generally, the colder the soil, the colder the squirrel and the longer the period of torpor.

During hibernation, the squirrels metabolize fat reserves built up during their active season. Most of this fat is used during revival periods between hibernation bouts when the squirrel rapidly warms up and stays warm for several hours. Thus, arousals are metabolically expensive. Males usually end their hibernation about a week before they appear above ground, while females end it the day before they appear above ground.

Table 9.2. Physiological rates of black bears and arctic ground squirrels.

	Summer respiratory rate	*Hibernating respiratory rate*
Black bear	30 breaths/minute	2 breaths/minute
Ground squirrel	60 breaths/minute	Holds breath for 30 minutes, takes 10–15 breaths, repeats
	Summer body temperature	*Hibernating body temperature*
Black bear	37°C	30°C
Ground squirrel	37°C	−2°C

HUMAN HIBERNATION

So, it seems we have a reasonable understanding of how animals hibernate, but how would you induce an extended comatose state in a human? Well, although the procedure is currently barely beyond the science fiction arena, the technology may be operational in time for the first outer-planets mission thanks to research efforts conducted by the European Space Agency (ESA). ESA scientists have found that human hibernation strategies will most likely mirror those of the arctic ground squirrel and the marmot – animals that undergo three defined periods of hibernation: entry, the hibernation period, and exit. Let's take a look at how humans would prepare for each of these stages.

Entry

First, crewmembers would be required to attain a very high level of fitness to maximize their body's ability to deal with the stress of hibernating and the deleterious effects of being in stasis for several months. After being launched into low Earth orbit (LEO), crewmembers would enter the *hibernaculum* – a highly advanced medical facility. Here, flight surgeons would connect astronauts to intravenous tubes, through which fluids and electrolytes would be administered to compensate for changes in blood composition during the hibernation. Then, administration of a hibernation-inducing compound (the HIT) would place the astronauts in a state of hibernation. The key to putting astronauts in a state of hibernation may lie in a synthetic, opioid-like compound called *Dadle*, or *Ala-(D) Leuenkephalin*, which, when injected into squirrels, can put them in a state of hibernation during the summer. This research has been already extended to studies investigating the effect of applying Dadle to cultures of human cells, revealing human cells divide more slowly when Dadle is applied. In conjunction with the studies investigating Dadle, researchers are testing compounds such as *dobutamine* and *insulin-growth factor* (IGF). Dobutamine is normally administered to bedridden patients to strengthen their heart muscles but, in the case of hibernating astronauts, the compound would be administered to maintain health during the long period of inactivity. IGF would be administered to boost the astronauts' immune systems, which would be depressed during the long period of inactivity.

Hibernation period

During the hibernation, a suite of medical sensing and hibernation administration facilities would monitor the state of the hibernating astronauts. In addition to ensuring body temperature, heart rate, brain activity, and respiration stayed within normal boundaries, the medical equipment would also monitor blood pressure, blood glucose levels, and blood gases (we'll talk more about how these parameters are monitored later). If the flight was to the outer planets, requiring a VASIMR-assisted transit time of more than 800 days, it's possible that crewmembers would be woken up periodically – a strategy practiced by many animal hibernators. These short periods of consciousness would allow crewmembers to catch up on their exercise routine and eat real food. However, because of the problems of entering and exiting hibernation, it is possible this strategy might be deemed too risky, although crews would be woken in the event of an emergency such as extreme solar flare activity.

Exactly what would happen to the astronauts as they hibernated isn't known. While humans have never had the need to hibernate for protection against the elements as some animals do, is it possible that we once had the biological mechanisms to regulate our metabolic activity and temperature for long periods of time? Do we still have those mechanisms and just not use them, perhaps? Until scientists perform more studies, we won't know, so the best answers we can come up with are based on similar states to hibernation such as meditation, sleep, and starvation. These states are characterized by many of the same variables as hibernation, such as decreased metabolic activity, decreased oxygen consumption, relaxed muscle, and decreased hormone production. Of the three states, meditation comes closest to hibernation, although a big difference between hibernation and meditation is that hibernating animals are not conscious during dormancy, while humans have demonstrated alpha-theta brain waves (those that are most closely related to being awake) during meditation. Animals are also, unlike humans, able to control the rate at which their bodies use lipids rather than carbohydrates for energy.

Exit

After their multi-year interplanetary trip, astronauts would be revived. This event would probably occur several weeks prior to entering orbit due to the deleterious effects of having been in hibernation for so long. But how would the long period in hibernation have affected the crew? Well, the truth is, no one knows. Let's see what Hollywood thinks might happen. Here's another scene from *Pandorum*, shortly after Bower has woken from hypersleep:

INT. STORAGE LOCKER TOOM -- MOMENTS LATER

BOWER walks along a row of storage lockers reading each name plate. He stops at the locket labeled; Corp. Bower. He pauses, mumbling the name once again with no recognition.

                         BOWER
     Bower.

He opens the locker to see an assortment of personalized belongings, uniform and gear, none of it seeming familiar.

He uses a towel to wipe off the layer of sweat and oily substance coating his body, pausing to see an identification tattoo on his forearm that reads: FLT>>005>>015.

He pulls on a black flight suit, emblems and military markings of rank on the chest. He pulls on shoes and a tool belt with a high-powered pen-lamp.

On the inside of the locker is an emergency manual card with block figures displaying breathing and stretching exercises, labeled; Hyper-Sleep Disorientation Recovery Procedures.

                         BOWER
                   (reading to himself)
... in the event of mild memory loss ... Resulting from extended —

He freezes on ... A PHOTO tucked inside the door. EVALON, a beautiful young woman smiling towards the camera. He stares at the photo, his frown deepening with confusion. He checks the back for a name, but nothing. He tucks the photo into a pocket and exits.

> Excerpt from the screenplay *Pandorum*,
> by Travis Milloy, *www.mypdfscripts.com*

In *Pandorum*, hypersleep leaves crewmembers with total amnesia, so Payton and Bower don't know who they are. Normally, there's supposed to be someone there to help re-orient them when they reawaken from hypersleep, but they're on their own. Once again, we just don't know how several years in stasis will affect astronauts. However, research has shown the deep torpor associated with hibernation may be problematic for the brain, so having a handbook outlining the disorientation recovery procedures will probably be helpful. The reason for the discombobulating effects of extended stasis occurs during torpor entry, when the body's temperature is gradually reduced. The cooling process results in reduced cortical power and profound differences in sleep architecture and memory consolidation. More worrying for those awakening from hibernation are the potentially deleterious effects upon spatial memory and operational conditioning. Imagine waking up from years in stasis (like Bower, in *Pandorum*) and not knowing where you are. Of course, until hibernation is performed on humans, we just won't know for sure. It's possible that humans will need only to follow a few instructions written in the Hypersleep Disorientation Recovery Procedures manual to fully recover from stasis, or it's possible that being comatose for several years might result in more insidious effects as portrayed in *Pandorum*.

INT. INFIRMARY
BOWER climbs through the debris towards the apparent window, but it's pretty far up a slanted wall and he begins to climb.

                         BOWER
I'm just thinking, maybe there's something we haven't considered. Eden failed ... because of ODS syndrome. The one officer who was suffering from Pandorum. Cabin fever. Maybe there was something similar with one of the earlier teams —

                         PAYTON
No, no, no. You're talking about — Eden was lost

because of a mechanical failure, Simple as that. A systems malfunction.

BOWER pauses his climb, looking up towards the 'apparent window', inside what appears to be a bluish moon, dotted with craters. He eagerly continues his climb, growing closer.

BOWER

That was the report . . . but don't you remember the story of the one flight officer . . . the real story? They say he went lunar, launched the entire ship, five thousand people killed with the flip of a switch. All from a bad case of cabin fever —?

PAYTON

Spook story, along with a million others. Eden was mechanical failure, not human error. A glitch . . . that was remedied way before we ever —

BOWER

And why did they switch the rotation shifts from three years to two? And they eliminated the manual over-ride —

PAYTON

They did that for obvious reasons, not because of Eden.

BOWER pauses . . . within reach of the window only to realize it was an illusion. A glass compartment door left slightly ajar was reflecting a circular light panel, smeared with oil to give the illusion of craters. His shoulders drop in defeat.

BOWER

Well, at least we didn't wake up . . . floating away in a coffin.

PAYTON
(under his breath)

Who says we didn't?

BOWER climbs down, peering out into the corridor.

BOWER

Something happened to the other team? Maybe there was . . . Pandorum.

Excerpt from the screenplay *Pandorum*, by Travis Milloy, www.mypdfscripts.com

STASIS MONITORING

In *Pandorum*, one of the flight crews is woken early – an event that would hopefully be prevented in the future by the use of an AI monitoring system. Such a system will require a full range of medical monitoring and administration equipment for each astronaut. The system would monitor medical parameters such as body temperature, electrocardiogram, heart rate, brain activity, gas exchange, blood pressure, and a whole host of other variables ranging from blood glucose and metabolite levels to gas analysis and clotting times. The information would be passed on to the AI stasis agent and to the ground station on Earth for analysis. The AI stasis agent would act as a nurse for the duration of the voyage. It would be capable of interpreting medical information supplied by the monitoring system and acting on that information in a timely manner. If a problem developed, the agent might consult with the ground station, depending on the communication delay. If it was judged that the contingency was serious and the communication delay too long, the agent would intervene autonomously.

The monitoring agent would probably be organized in two levels. The higher level would monitor fault detection, diagnosis, planning, and explanation, while the lower level would be responsible for perception, data acquisition, dealing with messages from flight surgeons at Mission Control, and communications with the ground. The agent would probably be loaded with a knowledge base organized under six organ systems (cardiovascular, pulmonary, renal, hematological, neurological, and metabolic/endocrine). This knowledge base would contain information on dozens of diseases and complications, hundreds of parameters, signs and symptoms, and all sorts of treatment actions and plans. As the voyage progressed, the agent would continually be updated with the existing knowledge base from the ground.

Because of the need to operate autonomously in the event of a medical emergency, the agent would be capable of three major reasoning components. The first of these would perform data analysis and interpretation, the second would perform diagnoses and therapy management, while the third would perform protocol-based treatment. A central monitoring computer (CMC) would be used as the core element of the sensor monitoring system. The CMC would be responsible for gathering data sent by all the medical sensors and logging and updating the data gathered in the central database. Each astronaut would wear a sensor unit that would monitor and transmit their vital signs to the CMC. The sensor unit would also receive commands from the CMC and respond appropriately.

Among its many monitoring tasks, the CMC would perform calculations on data being sent from mission control and determine what tasks to perform. These tasks could include requesting additional data from Mission Control, sending a command to the hibernaculum, or even adjusting environmental conditions. It would also be sensitive to predefined safety limits. For example, in the event of a radiation burst, the CMC would notify Mission Control and perhaps even initiate an alarm to wake the crew.

Life support

Thanks to the astronauts sleeping on their way to their destination, the requirement for LSSs will be scaled down, but the crew will still need life support for the hibernaculum and the common life-support functions such as air revitalization, waste processing, hygiene, fire detection and suppression, ventilation, and contamination control. Food production and preparation facilities will be redundant and hygiene facilities will obviously be scaled down and perhaps even eliminated altogether. There will also be a requirement for hydration due to the loss of moisture through respiration, but this function will probably be met by the hibernaculum control system. Waste processing will also be scaled down, as the wastes from the hibernaculum will mostly be particulates out-gassing from the crew and the stasis compartment. One aspect of life support that will need to be considered is the inclusion of an artificial gravity capability due to the prolonged immobility of the crew and it's possible the hibernaculum will actually be integrated into an artificial gravity facility to prevent the crew from losing too much bone and muscle.

One issue will be how to integrate the hibernaculum with the spacecraft's LSS. The life-support role will probably be split between the ship-dedicated LSS (which will be dormant for most of the voyage) and the hibernaculum's LSS. This arrangement will result in a two-phase LSS, the design of which would be determined by the hibernation strategies employed by mission planners. For example, if mission planners decided it was necessary to wake the astronauts occasionally, the primary LSS would have to be designed to be periodically reactivated. It might not sound like much of a problem, but biological LSSs can take several weeks to start up, so periodic waking probably won't be an option.

If thoughts of long-duration space journeys and hibernation conjure up images of the opening scene of *Alien*, you're not alone. Although placing astronauts into hibernation would solve many problems during the deep-space phases of an ECM, several issues remain unresolved. Scientists still need to develop a trigger (HIT) compound capable of inducing a state of hibernation and research concerning the secondary effects of hibernation is still lacking. For example, the effects of hibernation on memory, the metabolism, or the immune system are unknown. Another problem are the deleterious effects of zero gravity combined with the inactivity of hibernation, although this may be resolved by using some means of artificial gravity. Other challenges include problems associated with how the hibernation state is induced, established, regulated, and exited, and how to administrate compounds to a hibernating human. Achieving and perfecting human hibernation will require expertise in, and integration of, pharmacology, genetic engineering, environmental control, medical monitoring, AI, radiation shielding, therapeutics, spacecraft engineering, *and* life support. Only when *all* these disciplines have been successfully integrated will stasis be capable of making long-haul spaceflight a little more comfortable.

While the concept of stasis may seem futuristic, the daunting timeframe facing astronauts means hibernation is an idea that has to be taken seriously. Apart from the boredom of a lengthy transit, there are the powerful logistical reasons to place astronauts in hibernation. At the moment, the level of inquiry really is just speculative, but while stasis may seem the stuff of science fiction, as so often happens, science fiction has a habit of becoming fact.

Appendix: The Interplanetary Bioethics Manual

The Interplanetary Bioethics Manual (IBM) describes biomedical ethical principles to provide crew medical officers (CMOs) and Commanders with guidance in resolving ethical problems that may occur during exploration class missions (ECMs). The IBM is not a substitute for the experience and integrity of CMOs. The IBM is intended to facilitate the process of making ethical decisions in austere environments in which there is limited or no abort capability, limited life-support supplies, and restricted on-site medical support. The IBM presents general guidelines only. In applying these guidelines, CMOs and Commanders should consider the circumstances of the crewmember at issue and use their best judgment.

Medical ethics is based on the principles from which positive duties emerge. These principles include beneficence (a duty to promote good and act in the best interest of the patient and the health of society) and non-maleficence (the duty to do no harm to patients). The relative weight granted to these principles and the conflicts among them may account for the ethical dilemmas CMOs and Commanders may face during ECMs.

PATIENT

When medical capabilities permit, the CMO's primary commitment shall be in the crewmember's best interests, whether the CMO is preventing/treating illness or helping crewmembers cope with illness, disability, or death. The degree to which the interests of the crewmember will be promoted shall be determined by the life-support consumables and medical capabilities available.

At the beginning of and throughout the crewmember–CMO relationship, the CMO shall work towards an understanding of the crewmember's health problems, concerns, goals, and expectations. After the crewmember and CMO agree on the problem and the goal of therapy, the CMO shall present one or more courses of action. After consulting with ground-based flight surgeons, the CMO will initiate a course of action.

CONFIDENTIALITY

To uphold professionalism and protect crewmember privacy, CMOs shall limit discussion of care issues to the Commander and ground-based flight surgeons.

DISCLOSURE

To make health care decisions and work intelligently in partnership with the CMO, the crewmember, Commander, and Mission Control must be well informed. Information should be disclosed whenever it is considered material to the patient's understanding of his or her situation, possible treatments, and probable outcomes. This information includes the burdens of treatment, the nature of the illness, and potential treatments. Information that is essential to and desired by the crewmember must be disclosed.

Information should be given in terms that the patient can understand. The CMO should be sensitive to the patient's responses in setting the pace of communication, particularly if the illness is very serious. Disclosure and the communication of health information should never be a mechanical or perfunctory process. Upsetting news and information should be presented to the patient in a way that minimizes distress.

In addition, CMOs shall disclose to patients information concerning procedural or judgment errors made in the course of care if such information is material to the patient's well-being. Errors do not necessarily constitute improper, negligent, or unethical behavior, but failure to disclose them may.

INFORMED CONSENT

The patient's consent allows the CMO to provide care. Consent may be either expressed or implied. When the patient presents to the CMO for evaluation and care, consent can be presumed. The underlying condition and treatment options shall be explained to the patient and treatment shall be rendered or refused. In medical emergencies, consent to treatment that is necessary to maintain life or restore health is implied unless it is known that the patient would refuse the intervention.

The doctrine of informed consent goes beyond the question of whether consent was given for a treatment or intervention. Rather, it focuses on the content and process of consent. The CMO is required to provide enough information to allow a patient to make an informed judgment about how to proceed. The CMO's presentation shall be understandable to the patient and shall include the CMO's recommendation.

The principle and practice of informed consent rely on patients to ask questions when they are uncertain about the information they receive; to think carefully about their choices; and to be forthright with the CMO about their concerns and reservations about a particular course of action. Once a patient and the CMO decide on a course of action, the patient shall make every reasonable effort to carry out the aspects of care that are in their control.

The CMO is obligated to ensure that the patient is adequately informed about the nature of the patient's medical condition and the objectives of, alternatives to, possible outcomes of, and risks involved with a proposed treatment.

DECISION-MAKING CAPACITY

When a patient lacks decision-making capacity, the CMO, in consultation with Mission Control, shall make decisions on the patient's behalf. In these cases, CMOs shall refer to the patient's preferences and act in the best interests of the patient unless those interests compromise mission safety. CMOs, in consultation with Mission Control, shall take reasonable care to ensure decisions are consistent with those preferences and best interests. When possible, these decisions should be reached in consultation with flight surgeons and other physicians. If disagreements cannot be resolved, the final authority shall be the Commander.

DECISIONS ABOUT REPRODUCTION

In the event that a crewmember becomes pregnant and no abort-to-Earth capability is available or an abort to Earth would jeopardize the mission, the CMO has a duty to terminate the pregnancy.

CHRONIC, OVERWHELMING, AND/OR CATASTROPHIC ILLNESSES

In the event of a patient suffering a chronic, overwhelming, and/or catastrophic illness that places an excess demand on life-support consumables, the CMO shall euthanize the patient in accordance with mission guidelines.

PATIENTS NEAR THE END OF LIFE

Palliative care near the end of life shall not be administered if such care places undue demands on life-support and/or medical consumables. When circumstances permit, families of patients near the end of life shall be prepared for the course of illness and care options at the end of life. Ground-based clinicians should be able to assist family members and loved ones experiencing grief after the death of the patient.

MAKING DECISIONS NEAR THE END OF LIFE

Crewmembers with decision-making capacity have the legal and ethical right to refuse recommended life-sustaining medical treatments. The patient has this right regardless of whether he or she is terminally or irreversibly ill, has dependents, or is pregnant. The patient's right is based on mission safety. These situations demand empathy, thoughtful exploration of all possibilities, and, when operational circumstances permit, may require additional consultations.

Patients without decision-making capacity have the same rights concerning life-sustaining treatment decisions as mentally competent patients. Treatment shall conform to mission guidelines.

ADVANCE CARE PLANNING

Advance care planning allows a competent crewmember to indicate preferences for care and choose a surrogate – normally this will be the CMO – to act on his or her behalf in the event that he or she cannot make health care decisions. Advance planning shall comprise written advance directives, such as a living will for health care, which enables the patient to appoint a surrogate who will make decisions if the patient becomes unable to do so. The surrogate shall be obligated to act in accordance with the patient's previously expressed preferences or best interests of the mission. When there is no advance directive and the patient's values and preferences are unknown or unclear, decisions shall be based on the mission's best interests.

WITHDRAWING OR WITHHOLDING TREATMENT

Withdrawing and withholding treatment are equally justifiable, ethically and legally. Treatments should not be withheld because of the mistaken fear that if they are started, they cannot be withdrawn. This practice would deny patients potentially beneficial therapies. Instead, a time-limited trial of therapy could be used to clarify the patient's prognosis depending on life-support and medical consumables. At the end of the trial, a conference to review and revise the treatment plan shall be held.

DO-NOT-RESUSCITATE ORDERS

Intervention in the case of a cardiopulmonary arrest is inappropriate for some patients, particularly those with terminal irreversible illness whose death is expected and imminent. Because the onset of cardiopulmonary arrest does not permit deliberative decision-making, decisions about resuscitation must be made in advance.

Do-not-resuscitate orders or requests for no cardiopulmonary resuscitation shall specify care strategies and orders that describe all other changes in the treatment goals or plans. In the event that a resuscitation effort cannot conceivably restore circulation and breathing, the physician should help the family to understand and accept this position. The CMO who writes a unilateral do-not-resuscitate order must inform the patient.

Any decision about advance care planning, including a decision to forgo attempts at resuscitation, shall apply in every care setting for that patient. Decisions made in one setting shall consider future situations and the appropriateness of applying that decision in that setting. In general, a decision to forgo attempts at resuscitation should apply in every setting – spacecraft and planetary habitat.

DETERMINATION OF DEATH

The irreversible cessation of all functions of the entire brain is an accepted legal standard for determining death when the use of life support precludes reliance on traditional cardiopulmonary criteria. After a patient has been declared dead by brain-death criteria, medical support shall be discontinued.

IRREVERSIBLE LOSS OF CONSCIOUSNESS

Crewmembers who are in a persistent vegetative state are unconscious but are not brain dead. They lack awareness of their surroundings and the ability to respond purposefully to them. Because a persistent vegetative state is not itself progressive, the prognosis for these patients varies with cause. However, due to limited life-support and medical consumables, patients in a persistent vegetative state shall not be given life-prolonging treatment.

CMO-ASSISTED SUICIDE AND EUTHANASIA

Patients and CMOs may find it difficult at times to distinguish between the need for assistance in the dying process and the practice of assisting suicide.

CMO-assisted suicide occurs when the CMO provides a medical means for death, usually a prescription for a lethal amount of medication that the patient takes on his or her own. In euthanasia, the CMO directly and intentionally administers a substance to cause death. CMOs and patients shall distinguish between a decision by a patient or authorized surrogate to refuse life-sustaining treatment or an inadvertent death during an attempt to relieve suffering, from CMO-assisted suicide and euthanasia. Mission guidelines concerning moral objections to CMO-assisted suicide and euthanasia should not deter CMOs from honoring a decision to withhold or withdraw medical interventions in situations dictated by the mission.

In the mission setting, all of these acts must be framed within the larger context of the good of the mission. Many patients who request assisted suicide have uncontrolled pain, or have potentially reversible suffering. In such cases, the CMO may withdraw life support and/or increase medication to shorten life.

Appendix

OBLIGATIONS OF THE CMO TO THE CREW AND TO THE MISSION

All CMOs must fulfill the profession's collective responsibility to advocate the health and well-being of the crew.

Index

Abortion, 125, 126
Advance Health Care Directive, 120
Advanced Life Support, 20
Advanced Multiple-Projection Dual-energy X-ray Absorptiometry, 81
Advanced Resistive Exercise Device, 10
Amundsen, Roald, 43, 112
Aquarius, 146, 147
Asthenia, 107
Australian National Antarctic Research Expeditions, 29, 30

Barratt, Michael, 21
BioSuit, 155
  Mechanical counter-pressure, 155, 156
Bone mineral density, 77

Cell rover, 64
Centisievert, 60
Combined Operational Load-Bearing External Resistance Treadmill, 11, 83, 84
Computed tomography, 150
Concordia, 102, 103
Consultancy telemedicine, 141
Coronal mass ejection, 58
Crew Medical Officer, 15, 19, 20, 21, 22, 23, 26, 27, 30, 31, 32, 33, 34, 35, 125, 139, 141, 142, 145, 148, 152
Crew Resource Management, 109

da Vinci Surgical System, 144
Dadle, 166
Decompression Sickness, 19

Defense Advanced Research Projects Agency, 153
Definitive Medical Care Facility, 17, 20
Dendrimers, 69
Dobutamine, 166
Dual-Energy Absorptiometry, 80

Eicosapentaenoic acid, 83
Eleutherococcus Senticosus, 31
Euthanasia, 127
Extracorporeal shock wave lithotripsy, 24

Five Wishes, 119, 120
Fosamax, 82
Freitas, Robert, 71

Galactic cosmic rays, 5, 66
Gattaca, 48, 131–133

Health Care Power of Attorney, 120
Hibernaculum, 163, 166, 171
Hibernation induction trigger, 163, 166, 171
Human Outer Planets Exploration, 88

Innerspace, 150, 151
Insulin-growth factor, 166
Interactive telepresence, 142
Interplanetary Bioethics Manual, 26,
International Commission on Radiological Protection, 65

Isolation Study for European Manned Space Infrastructure, 103

Joosten, Kent, 89, 90

K2, 130, 131
Kidney stones, 24
Kohler, Pierre, 121

Lapierre, Judith, 104
Linear energy transfer, 65
Line-Oriented Flight Training, 110

Magnetic Resonance Imaging, 80
Matroshka, 6, 7, 8
Mawson, Douglas, 43
Medical risks, 4
Miacalcin, 82
Million Clinical Multiphasic Inventory, 44
Minnesota Multiphasic Personality Inventory, 44
Mission Medical Pack, 23
Moon (movie), 107, 108, 110, 111, 134–137
Multilateral Space Medicine Board, 25
Mutation (DNA), 63

Nanocomputers, 152
Nanocapsules, 152
Nanomedibots, 152
Nanoparticles, 69, 151
Nansen, Fridtjof, 43, 96, 97, 99, 104, 105, 112
NASA Extreme Environment Mission Operations, 145, 146, 148
NASA Institute for Advanced Concepts, 88, 155
National Council on Radiation Protection, 65
National Space Biomedical Research Institute, 107
Newman, Dava, 155
Nonmaleficence, 119
Nowak, Lisa, 45, 46, 109, 124
Nuclear track detector package, 8

Oates, Captain, 128
Osteoporosis, 78, 79

Pandorum, 107, 159, 167–169
Phoenix, Christopher, 71

Private Medical Conference, 22, 25, 124
Problem-Oriented Team Supervision, 110

Raven IV, 147, 148, 149
Relative Biological Effectiveness, 60
Rem, 60
Remote monitoring (telemedicine), 142
Richardson ground squirrel, 165
Rogozov, Leonid, 49–52
Romanenko, Yuri, 31, 86

Shackleton, Ernest, 43, 44, 94, 99, 101, 104, 105, 109, 111, 112, 130
Short-radius centrifuge, 87
Sievert, 60
Simulation of Flight of International Crew on Space Station, 103, 104
Skylab-4, 11, 102, 105, 106, 109
Slow Release Sodium Fluoride, 82
Solar flares, 58
Solaris, 107
Solar Particle Events, 57, 66
Space Radiation Analysis Group, 30
Space Radiation Laboratory, 64
Space trauma pod. 154
  Medical encounter record, 154
  Patient imaging system, 154
  Patient registration subsystem, 154
  Scrub nurse subsystem, 155
  Supply dispensing subsystem, 153, 154, 155
Store-and-forward telemedicine, 142

Telesurgery, 142
Telementoring, 143
Thermo-luminescence dosimeter, 8
Tissue-engineered Organ Replacement System, 34

Vasculocyte, 72, 74
Virtual Space Station, 107, 108

Williams, Sunita, 85
Wolf, David, 47, 86
Wolpe, Paul, 121
Wyle Laboratories, 87

GPSR Compliance

The European Union's (EU) General Product Safety Regulation (GPSR) is a set of rules that requires consumer products to be safe and our obligations to ensure this.

If you have any concerns about our products, you can contact us on

ProductSafety@springernature.com

In case Publisher is established outside the EU, the EU authorized representative is:

Springer Nature Customer Service Center GmbH
Europaplatz 3
69115 Heidelberg, Germany